LUFTWAFFE CAMOUFLAGE AND MARKINGS 1933-1945

Photo Archive 1

LUFTWAFFE CAMOUFLAGE AND MARKINGS 1933-1945

Photo Archive 1

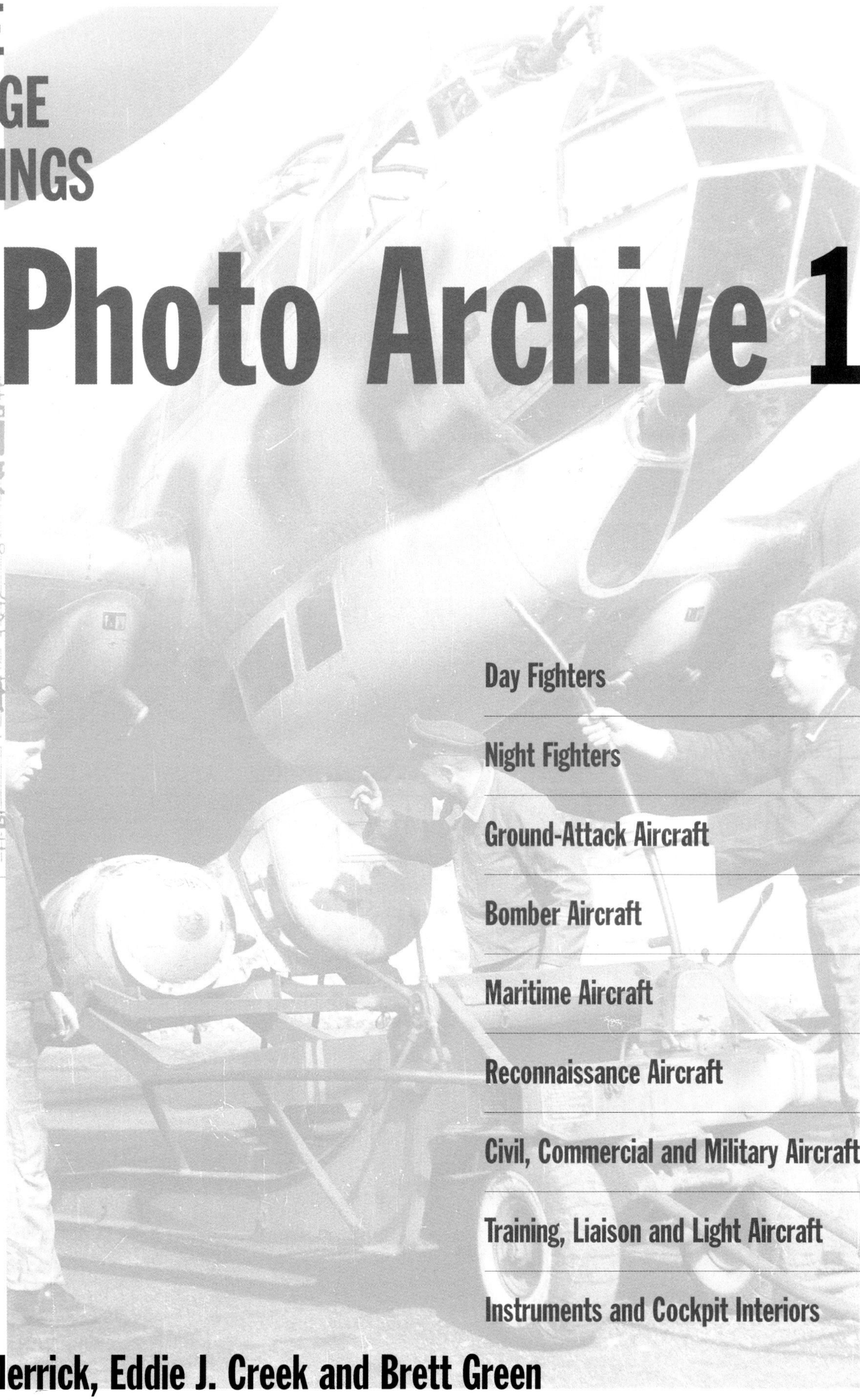

Day Fighters

Night Fighters

Ground-Attack Aircraft

Bomber Aircraft

Maritime Aircraft

Reconnaissance Aircraft

Civil, Commercial and Military Aircraft

Training, Liaison and Light Aircraft

Instruments and Cockpit Interiors

MIDLAND

An imprint of
Ian Allan Publishing

K.A. Merrick, Eddie J. Creek and Brett Green

THE AUTHORS

Ken Merrick

Ken Merrick has been researching and writing about British and German aviation subjects for the last 40 years, producing books ranging from the history of individual aircraft types, through to a highly detailed history of Allied clandestine air operations in World War Two. He became particularly interested in the subject of German camouflage systems some 30 years ago and has since written a number of highly acclaimed books on the subject in the intervening years including *Luftwaffe Camouflage and Markings 1935-1945 Volume One* (1973), *German Aircraft Markings 1939-1945* (published by Ian Allan in 1977), *The Official Monogram Painting Guide for German Aircraft 1935-1945* (with Thomas H. Hitchcock, 1980) and *German Aircraft Interiors 1935-1945 Volume One* (1998).

This work is a compilation of all those years of research, melded with intriguing new information culled from a mass of official RLM and German paint industry documentation. The result is a complete and radical re-appraisal of all that has appeared on this contentious subject during the intervening years. Ken lives in Adelaide, South Australia.

Eddie Creek

Eddie Creek has been involved in studying the history of the Luftwaffe for more than forty years and during this time has been privileged to have met many WW2 veterans who were kind enough to relate their experiences, and often pass over their photographs and documents. In the early years he was also extremely lucky in meeting many of the older generation of historians who, together with most of the veterans, have now sadly virtually all passed away. However, their help and gifts have helped him to collect a substantial archive of material which he is pleased to be able to share with the readers who buy the books he has been involved in. Eddie lives in the South East of England.

Brett Green

Brett Green is the Editor of the modelling websites HyperScale **(www.hyperscale.com)** and Missing-Lynx **(www.missing-lynx.com)**. He is author of two previous books on Luftwaffe camouflage and markings, 'Augsburg's Last Eagles' and 'Stormbird Colors', as well as seven titles in the Osprey Modelling and Masterclass series. In addition to his enthusiasm for aviation colours, Brett also has a keen interest in scale models, photography and flying. He lives in Sydney, Australia with his wife and their two children.

Acknowledgements

The following persons and institutions are gratefully thanked for the use of their photographs, either reproduced in this volume or as reference material for preparation of captions.

W. Berge, R. Chapman, M. Clayton, S. Coates, D. Cormack, P. Cohausz, E. Creek, J. Crow, M. Drewes, W. Falk, F. Grabowski, B. Hay, R. Harris, W. Hesze, O. Hintze, M. Jessen, H. Jonas, B. Ketley, K. Kössler, R. Langsford, R. Lutz, H. Mankau, H. Nowarra, H. Obert, M. Pegg, P. Petrick, T. Poruba, M. Ullmann, J. Radziwill, B. Rosch, S. Santos, B. Sartwell, F. Sellinger, F. Smith, E. Sommer, L. Tabori, H. Thiele, K. Tiedemann, G. Valentini, J. Vasco, D. Vincent, D. Wadman, Australian War Memorial, Hikoki Publications, JaPo Publications, Lufthansa, Royal Danish Air Force and the Smithsonian Institution.

Luftwaffe Camouflage and Markings 1933-1945,
Photo Archive 1
© 2007

ISBN (10) 1 85780 275 6
ISBN (13) 978 1 85780 275 7

Produced by Chevron Publishing Limited
Project Editors: Chevron Publishing Limited
Cover and book design: Mark Nelson

Published by Midland Publishing
4 Watling Drive, Hinckley, LE10 3EY, England
Tel: 01455 254 490 Fax: 01455 254 495
E-mail: midlandbooks@compuserve.com

Midland Publishing is an imprint of Ian Allan Publishing Ltd
Worldwide distribution (except North America):
Midland Counties Publications
4 Watling Drive, Hinckley, LE10 3EY, England
Telephone: 01455 254 450 Fax: 01455 233 737
E-mail: midlandbooks@compuserve.com
www.midlandcountiessuperstore.com

North American trade distribution:
Specialty Press Publishers & Wholesalers Inc.
39966 Grand Avenue, North Branch, MN 55056, USA
Tel: 651 277 1400 Fax: 651 277 1203
Toll free telephone: 800 895 4585
www.specialtypress.com

Printed in England by Ian Allan Printing Ltd
Riverdene Business Park, Molesey Road,
Hersham, Surrey, KT12 4RG

Visit the Ian Allan Publishing website at:
www.ianallanpublishing.com

MIDLAND
An imprint of
Ian Allan Publishing

Contents

Introduction

Photographic albums can offer a mine of information for the historian, restoration teams and modellers, or those who simply enjoy examining the photographic record. No photograph should be simply taken at face value if the observer wishes to glean as much information as practicable from it. Photographic quality is the primary factor. Old, poorly exposed or badly printed photographs can keep information back, but with today's computer software programmes, the average computer-equipped person can now gain considerably

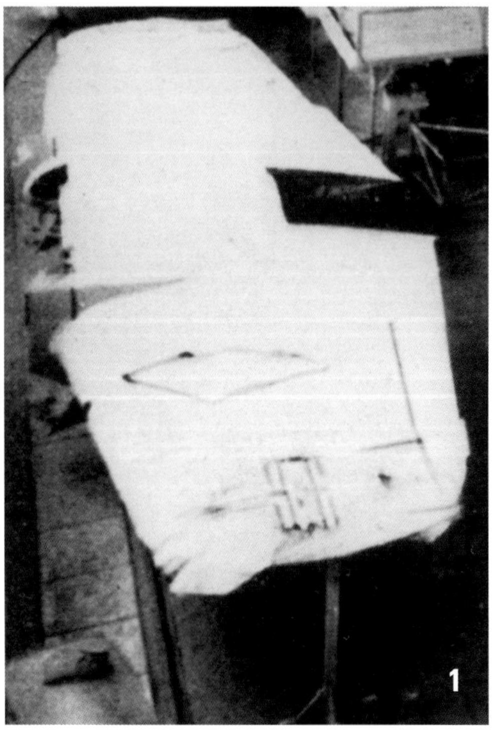

more information from even those damaged photographic resources.

This photograph showing Dornier Do 17s under construction will serve as an example of what can be deduced from what is, photographically, an average quality image. The overall view simply shows Do 17s in various stages of assembly, from which it is possible to trace the assembly sequence. (See image above).

On the left-hand side of the factory the Balkenkreuz is being applied to wings, previously painted in RLM 02, using masking tape and brown paper.The white sections of the marking have been sprayed on, covering the protective paper with overspray. (See image number 1).

To the right, full tailplane assemblies are being attached to the fuselage structure. Like the wings, these units have had their horizontal tail surfaces pre-painted in RLM 02, but the fin and rudder units have two distinct colours on them. The bottom section of each fin is painted with 02 but the remainder have had a second coat

2

applied over a darker colour.This is the red of the fin band marking, the first element applied before the white disc and Hakenkreuz. Due to the size of that marking and the size of the rudder and fin, it was usual to pre-paint the entire rudder and the fin, from the bottom line of the red fin band upwards, in red. The small portions above and below the banner marking on the rudder and upper section of the fin were then painted over in the appropriate camouflage colour. This used more red paint but saved on time - something the German aircraft industry could certainly afford before the war. (See image number 2).

3

In the centre, and towards the back of the assembly room, the reason for that overpainting becomes clear. Of the two fin and rudder assemblies just visible, the one farthest away retains the old red banner, white disc and Hakenkreuz, while the one nearest has had the red band and white disc painted out, leaving the plain Hakenkreuz in its existing (now incorrect) position. (See image number 3).

Further to the right a completely camouflaged tail unit has the correct form of the revised Hakenkreuz, but again still in the wrong position. (See image number 4).

All of this helps to date the photograph as January 1939, the RLM having issued instruction Az. 19 d 10 Genst. 2.Abt. (III B) on 15 December 1939, with an effective implementation date of 1 January 1939 which required removal of the old-style marking and its replacement with a plain Hakenkreuz placed on the fin on a camouflaged background (see 'Luftwaffe Camouflage and Markings 1933-1945', Volume Two, page 436). The fact that the fin and rudder banner markings had been reduced to just the Hakenkreuz, but that this was left centrally placed, indicates that this scene was captured very close to the promulgation date of the new instruction - aircraft were under assembly when the order came into effect and thus the transition was captured on film. Given the rate of Do 17 production, one might expect the image to show the production stages that encompassed a few days at best. It is rare that a single photograph captures the moment of such a transition in Luftwaffe events.

Other minor details can be gleaned from the photograph, such as one of the fuselage units close to the right-hand side windows with its fuselage Balkenkreuz masked out with tape and paper; the structure of the mobile assembly platforms; the mating of fully painted and unpainted components; and so forth.

Photographs can provide hours of pleasure, as well as information. This collection has been assembled to amplify the wide-ranging selection contained in Volumes One and Two of 'Luftwaffe Camouflage and Markings 1933-1945' and some photographic captions have been cross referenced to those two volumes for the assistance of readers seeking more information. It can, of course, be enjoyed simply for what it contains; but whatever the focus of the reader we hope that this book brings pleasure and information to all.

The authors.

Foreword

The German war machine was rebuilt during the 1930s in defiance of the terms of the Treaty of Versailles. At the close of the First World War, Germany had been forbidden to operate military aircraft. Even so, secret preparations for a new air force were underway well before Hitler's rise to power.

As early as the 1920s, German youth was encouraged to participate in sports gliding and flying light aircraft, laying the foundations for a new generation of military pilots. Germany also enlisted the help of the Soviet Union with the establishment of a secret facility at Lipetsk in 1924 where German, Dutch and Soviet flyers trained in aircraft banned by the Treaty. Pilots also gained experience in civil airliners which would eventually be developed into bombers and military transports in the coming years. Germany's enthusiastic participation in, and organization of, international air racing competitions might have seemed at the time a tentative portent for a peaceful future. In reality, it should have been taken as a sinister omen because, in a few short years, many of the competing aircraft found themselves armed in the skies over Spain, or as advanced trainers for military pilots who would spearhead the Blitzkrieg of 1939 and 1940.

The formation of the *Luftwaffe* in February 1935 stripped away any pretence of conforming to the Treaty of Versailles. Development of military aircraft and training of personnel proceeded swiftly. This fledgling air force did not have to wait long to test its new tools and tactics. In 1936, Germany sent a group of "volunteers", complete with aircraft, crews and armament, to support the anti-Republican government revolt led by Francisco Franco in Spain. Many of the Legion Condor's aircraft and pilots who were destined to become household names in the Second World War were tested here.

By 1939, the Luftwaffe was still relatively small, but its aircraft were modern and its best pilots already proven in battle.

In the prelude to the *Luftwaffe*, German aircraft were not camouflaged. In the earliest years of the new air arm, aircraft were painted in protective coatings of grey or, in some cases, remained in natural metal, a carry-over from their earlier civil guise. Some aircraft were finished in bright livery of stripes and bands to easily identify particular units. By September 1935, the unmistakable red, white and black National Socialist flag made its first appearance across the fin of *Luftwaffe* aircraft.

Circumstances dictated a change of painting policy during the Legion Condor's deployment to Spain. Pale greys gave way to darker shades and disruptive patterns better suited to disguising fighter and bomber aircraft in the air and on the ground. By the beginning of the Second World War, camouflage for fighters, bombers and transport aircraft had standardized as two dark shades of green covering the uppersurfaces and the fuselage sides, with light blue below. These colours were ideal over the dense forests of Eastern Europe, and were also helpful in a defensive sense for hiding aircraft under foliage.

The rapid progress of the war into the more open farmlands of Western Europe and across the English Channel dictated another serious rethink. As Germany's war expanded into ever more diverse environments - the harsh winters of the Eastern Front, the endless sands of North Africa and the grey expanses of the North Atlantic - camouflage was adapted as required, if not always promptly. By the last chaotic year of the war, a combination of new policies and expedience saw significant variation in *Luftwaffe* camouflage.

German aircraft markings developed in parallel. Priorities and systems changed, with many interesting recognition systems introduced at different periods.

This book builds on Ken Merrick's detailed work in Volumes 1 and 2 of 'Luftwaffe Camouflage and Markings 1933-1945' by examining photographic case studies of German aircraft stretching from the fledgling, pre-Luftwaffe period until the last days of the Second World War. The evolution from civilian colours and markings to full military livery is thoroughly chronicled in pictures and words.

These photographs and their detailed captions illustrate not only the strict application of official Luftwaffe camouflage and markings, but also many of the fascinating exceptions seen on individual aircraft.

Ken Merrick, Eddie Creek and Brett Green

Day Fighters

Three of the nine A-0 variant He 51 A aircraft which were evaluated by the DVL Reklamestaffel Mitteldeutschland for the, as yet, clandestine Luftwaffe; D-IQEE (A-02), D-IJAY (A-05) and D-IDIE (A-06). Overall finish was RLM 02 with black civil registration lettering. RLM 63 would replace RLM 02 colouring for fighter aircraft with the introduction of the four-colour system of 61, 62, 63 and 65 in 1936.

A photograph taken on 20 April 1935 to record the ceremony during which JG 134 received its honorary title of Horst Wessel. Close examination of the original photograph reveals an anomaly amidst the line-up of Ar 65s on the right. Hitler's Ju 52 personal aircraft was civil coded D-2600, by special concession, and in the centre row of Ar 65 aircraft here, the fifth machine in from the left, also retained its old form of registration – D-28??. This was more than a year after the change to the registration system, which provides an interesting insight into how long some overdue changes could linger. The size of the registration marking of the revised civil codes also varied as can be seen between the He 51s in the front row and the Ar 65s in the other two rows.

In this line-up of He 51s from all three Staffeln of I./JG 135 at Bad Aibling, the bold unit emblem painted beneath the cockpit provides an eclectic mixture of markings information. All wear one of the 'traditional colours', (RLM 24 blue in this instance) bestowed on specific fighter-equipped units from 1 September 1936 (see Volume One, pages 125-126). The overall colouring was RLM 02, with the blue colour applied to the entire engine cowling, extending along the upper decking of the fuselage and used as a band around the rear end of the fuselage. The nearest aircraft, 'White 7' W.Nr. 885, had the 3. Staffel white disc identification marking repeated on either side of the engine cowling and on top of it; on either side of the fuselage on the band of blue; and diametrically above and beneath the fuselage just aft of the cockpit (see Volume Two, pages 242-245). The white disc marking on the rear section of the fuselage contained a simple black and white figure – a rarity for the times. The second aircraft, 'White 4', was from 1. Staffel, which carried no additional white markings or band extension around the rear of the fuselage. Next in line is 'White 1' of 2. Staffel as signified by the white band visible around the engine cowling (and repeated through the middle of the colour band around the rear of the fuselage); then another 'White 1', this time of 1. Staffel, which also had a dark colour to its spinner tip. Next was a 2. Staffel machine, then one from 3. Staffel, another from 2. Staffel, and then a 3. Staffel machine. This impressive flight line was probably posed for an official photographer, propaganda photographs being of great importance at this time to try to convince the world that the Luftwaffe was larger than it actually was.

An interesting grouping of Bf 109 D-1s with some, in the far distance, wearing what appears to be the ZG 2 emblem. On its own, in mid-picture, 'Red 2' has what appears to be a double white band around the rear of the fuselage, but closer examination reveals a dark central stripe, which could be red, the Staffel colour. Nearer, on the left, is 'Yellow 12', which has two yellow bands thinly edged in black. The use of yellow may indicate that these were 3. Staffel machines, possibly also from ZG 2. (see Volume Two, page 263)

The marking on this Ar 68 E identifies the pilot as the Gruppenstab Kettenführer of 1. Kette. This part of the Stab-system of markings, introduced in 1935 with the He 51, carried through when the Bf 109 began to replace the biplane force.

Yet another example of a unique form of band marking applied to a Bf 109. 'Red 1' was the mount of the Staffelkapitän of 2./JG 27, and carried a narrow red band edged thinly in white, applied at 45°, which continued right around the fuselage. The name on the cowling, "SAMOA", was applied in black. This was one of twelve aircraft delivered to the unit, each bearing the name of a former German colony.

There has often been speculation about the use of a single uppersurface colour for Bf 109s, using one of the two standard greens. The oversize style of Balkenkreuz on the wing of this Bf 109 E, 'White 2', dates to the period between October 1939 and early 1940, the actual date in this instance being October 1939 and the location Fliegerhorst Weil. The marking, a personal one, was that of the pilot, Lt. Troha of 7./JG 26. The camouflage appears to be monochromatic, but close examination of the wing root section (particularly the lowered flap), reveals the two standard colours of 70 and 71. The effects of the lighting angle as well as the viewing angle can make colour perception deceptive. (M. Ullmann)

An interesting, and at times enigmatic view of a mixed batch of Bf 109 Es at Messerschmitt's Augsburg plant, destined both for the Luftwaffe as well as the Swiss Air Force. The latter were part of an order for 80 Bf 109 E-3s, the earliest deliveries of which had commenced in April 1939. Several of the Swiss order carried civil registrations - D-IYUI (eventually to become J-365, D-IYUM and D-IYUS (J-369). The Bf 109 in the lower right foreground had already had its camouflage changed to Swiss military standard, and national military markings for that country applied to the wings. The white disc marking across fin and rudder however was part of the original Reichs und Nationalflagge marking, minus the Hakenkreuz; the latter would have been required pre-war had this aircraft been intended for delivery with a civil registration. (The white disc portion of the Reichs und Nationalflagge marking can just be seen on the tip of the tailplane of D-IYUS. This process of change to the Swiss machines probably occurred as a result of the opening of the war, the export niceties and accompanying civil markings having thus become redundant.) In the background, aircraft leaving the assembly line are undergoing their final checks before the routine test flight process. The machine at centre right, an E-3, still retains its Stammkennzeichen, commencing with BY and has had its Balkenkreuz marking revised to the new proportions introduced on 24 October 1939, but retained the old, 660 mm smaller overall dimensions (a common fault on many aircraft types during the initial change-over (See Volume Two, page 436). The aircraft's Werknummer 2351, was still visible above the Balkenkreuz over the 65-colouring.

Personal names occasionally were applied to the cowling area forward of the wing on some fighter aircraft, though the more usual location was below or near the cockpit. Photographed around April-May 1940, 'Fortuna' was flown by the Gruppenkommodore, Hptm. Kienik. Note the red and white-painted wheel chocks, colours used as a visual warning on a variety of aircraft items from retractable steps to (in later times) radar aerials. The bottom of the leading edge of each chock was equipped with metal teeth to grip the ground. (M. Ullmann)

The Bf 109 force was supplemented by the heavy fighters of the Bf 110-equipped units. However, unlike the Bf 109s, these heavy fighters retained their 70/71/65 camouflage, as shown on Bf 110 C-1 of 13.(Z)/LG 1, L1+IH, with uppersurface demarcation taken right down to the bottom of the fuselage. (J. Vasco)

As air combat over England intensified, many units operating the Bf 109 E started to tone down the all too bright 65-colouring of the side surfaces with a variety of colours and mottles (see, Volume One, pages 135-136). The official standard had been changed from 70/71/65 to 71/02/65 in December 1939 (see, Volume One, page 77) but older aircraft, already in service, were left in their existing 70/71/65 camouflage pending their next scheduled major servicing and overhaul period. This Bf 109 E-1, or E-3, wearing the emblem of 3./JG 54, exhibits some interesting variations. The position of the Hakenkreuz across both fin and rudder reveals that it was a pre-January 1939 production aircraft. As such, it would have retained its original 70/71 camouflage. Evidence of this may be seen by the narrow horizontal dark band of the original colour, on the fin, behind the Werknummer. The Balkenkreuz was of the pre-24 October 1939 size of 660 mm (see Volume Two, pages 436 and 439), further evidence that this was an old machine, possibly a reserve brought out by the pressure caused by losses during the air fighting of that summer of 1940. While many examples of Bf 109 Es wearing 660 mm size markings were recorded, they had had the proportions altered to the revised standard, with wider white outlines (yet another problem that had eventuated with the revision to existing markings). Some units, however, had taken their own initiative when the fighting became intense and widened the black edging to reduce the white outlines, as part of the desperate attempt to increase the camouflage qualities. This machine appears to have received that treatment at some stage because a ragged fringe of black paint can be seen around parts of the Balkenkreuz. It is too dark to have been 70, as comparison with the propeller blades shows. The addition of 02 to the fuselage sides was applied quite roughly and thinly in places, meeting the uppersurface 71/02 pattern along the spine of the aircraft in a ragged line, suggesting that this had been done at unit level. The Bf 109 E-3 in the background had been treated in a similar manner. The absence of any Staffel, individual or even tactical markings seems to indicate that the aircraft in the foreground had only recently been repainted – note the lack of exhaust staining. The presence of the second emblem, the well known pistol-wielding Mickey Mouse emblem used originally by 3.J/88 of the Legion Condor, suggests that the pilot may have been a veteran of that unit. Alternatively, it may have related to the pilot or aircraft having been transferred from 1./JG 51, which was using that emblem during this period.

Bf 109 kill markings during the 1940 period varied, with some units favouring simple vertical bars on the tail section painted in black, red or white. Others adopted different styles, such as shown here on Hptm. Balthasar's Bf 109 E-4, W.Nr. 1559. JG 3 chose to identify ground and air kills by the addition of an arrowhead, pointing in the appropriate direction, with some details added in white to the black bar. When the tactical colouring of 04 yellow had been applied to the rudder, a strip of the original 65-colour, bearing the kill markings, was retained. Note the discolouration where repair patches cover bullet damage on the horizontal tail surfaces. The fabric patch covering the access hole in the fin was also missing.

The dark 70/71 colouring on this Bf 110 B of II./ZG 26 is relieved only by the stylish, personal marking of a lightning flash in a colour of similar tone to that of the 65-colouring of the undersurfaces. The all-black code of 3U+BC is not easy to distinguish on this dark camouflage, which is why it was more normal practice to paint, or at least to outline, the aircraft's individual letter in the Staffel colour. (J. Vasco)

The effect of painting the aircraft letter in its appropriate Staffel colour is clearly illustrated here on 3U+AP, W.Nr. 3044, and 3U+GN. Both also accord with the RLM standard of repeating the aircraft letter beneath each wing tip. 'A' had been transferred, with its pilot, when Oblt. Rossiwal moved from 5. Staffel to take over as Staffelkapitän of 6./ZG 26, hence retention of the yellow Staffel colour to the tip of the spinners. Bf 110 C-4 3U+GN, W.Nr. 3536, in the background, was shot down over the Thames Estuary on 2 September 1940, its crew becoming Prisoners of War. Camouflage and national markings on both aircraft were regulation perfect. Marking of the Werknummer on the lower part of the rear section of the fuselage was normal for most Bf 110s of this period. Note that the individual letter on the uppersurface of the wing of the aircraft, from which the photograph had been taken had been overpainted. (J. Vasco)

A Bf 110 of II./ZG 1 illustrates the strong contrast between the dark greens of the uppersurface and the pale blue of the lower ones, which suited ground camouflage needs but did not work well for air fighting situations. The II. Gruppe badge, of three yellow wasps with white wings was the slightly simplified version, the original having a four-humped white cloud mass beneath the wasps, with the white faded out as it descended. This version, with two stylised cloud humps in white, became the norm. The spinner tips were in a dark colour, possibly green or blue rather than red, indicating a Stab machine. (J. Vasco)

Revised camouflage colours for the Bf 110s were introduced in May 1940 (see Volume One, pages 55 and 65), the very dark existing 70/71/65 scheme being replaced with the 71/02/65 colouring already in use on the Bf 109 force – and for the same reasons. The Zerstörer force was about to enter the anticipated heavy air fighting over Britain, and Luftwaffe aircraft at their French and Belgian bases were unlikely to suffer air attacks on any large scale, making ground defensive camouflage of far less importance to that needed for air-to-air fighting. This Bf 110 D-3, VC+E(F?)K, provides a good example of the revised scheme, 02 being used as a solid colour (behind the 'VC') or in combination with 71 as a dense mottle to tone down the side surface colouring of 65. The fins and rudders feature very eclectic treatment: closely spaced short strokes of both colours were used on the outer face of the starboard fin and rudder, while the inboard face of the port fin had a solid base colour of 02 with large heavy mottles of 71. The rudder, in contrast, has mottling in both colours over the 65-colouring. The nature of and reason for the horizontal patch of colour on the starboard fin surface is unknown, possibly a repair still in 65-colouring. (J. Vasco)

An interesting photograph of two 3./ZG 26 Bf 110 Cs, that span the transition in camouflage. Nearest is U8+CL, in the old 70/71/65 finish, while U8+FL, in the background, has the revised 71/02/65 scheme. Both had their individual Staffel letter painted in yellow, with the Gruppe colour of white on the spinner tip of the nearer aircraft. The latter also retains an earlier form of marking, a narrow white band around the rear section of the fuselage. The white nose section resulted from the change in tactical marking colour from yellow to white, introduced on 2 September 1940 (see Volume Two, page 267), and extended to the Bf 110 force for the first time. The paint was temporary, removable with benzine, but applied in most instances to the Bf 110s with a brush, sometimes resulting in poor coverage, as seen here. This was not a result of decay over time as U8+CL was lost on 6 September, just four days after being painted. (J. Vasco)

This air-to-air view of 3U+LR of 7./ZG 26 illustrates again the poor coverage afforded to some of the Bf 110s when the white tactical marking was added to the nose section. The aircraft letter 'L' was marked in yellow with the Gruppe colour of white added to the spinner tip. The Hakenkreuz was not always easy to discern against the new camouflage of 71/02. (J. Vasco)

Another air-to-air view, this time of U8+GL, W.Nr. 3353 of 3./ZG 26, showing the revised camouflage of 71/02/65, but retaining the narrow white band marking seen on the Bf 110s of the unit still wearing the old 70/71/65 scheme. This was an early form of tactical marking noted on Bf 110s and Ju 87s from the start of the war. The loss date for this aircraft – 27 September 1940 – combined with the revised camouflage introduced in May, gives a time frame for the extended use of that form of marking. In this instance the white nose marking appears to have been applied more substantially than on other aircraft of the Staffel. Note that while the aircraft letter 'G' was correctly applied in the Staffel colour of yellow, and spinners in the Gruppe colour of white, repeat of the individual letter beneath the wing had been overpainted. In the foreground the marking of the aircraft letter on the aircraft used to take the photograph had also been overpainted. While some aircraft retained the wing upper- surface marking – Bf 110s, Ju 87s and some Do 17s – most had eliminated it by the end of 1940 as it was thought to compromise camouflage. (J. Vasco)

Retention of the II./ZG 76 Shark's Mouth emblem reduced the area of white tactical marking, and its Bf 110s appear to have had the white most carefully applied only to the removable upper panel of the nose section, over the gun bay. The quality of the finish, compared with that of other Bf 110s so marked, is significant. Some of this unit's aircraft would take part in the operations against Yugoslavia and Greece, the opening phase of the Balkan campaign on 6 April 1941, still retaining the white marking. During the fighting 4./ZG 26 was hastily moved to Mosul in support of Rashid Ali's uprising in Iraq, which had started on 3 April. The speed of the transfer entailed removing the German national markings and replacing them with those of Iraq, the white marking being retained. Close-up detail shows the finish to be in excellent condition, indicative that permanent paint had been used, rather than the temporary lacquer mentioned earlier (see Volume Two, page 275). The fact that the unit had chosen to use a removable panel, which enabled spray painting to be used rather than application by brush, may have influenced the choice of lacquer used. Camouflage was 70/71/65, with the code 2N+D(?)N marked in black, the 5. Staffel colour of red not having been used for the aircraft letter, although the II. Gruppe colour (again red) had been used on the spinner tips. The pale colour underneath the nose section, immediately aft of the Shark's Mouth, is interesting as it stands out from the 65-colouring of the other undersurfaces. Yellow was not used on the Bf 110s during the air fighting over Britain. It may just have been fresh 65-colouring. There are at least three white Abschuß markings on the tip of the port fin. (J. Vasco)

The Bf 110s of 4./ZG 76 operated from Mosul in Iraq, in support of the April 1941 uprising led by Rashid Ali (see Volume Two, page 273). The aircraft were transferred from Greece where they had just taken part in Operation Marita. German nationality markings and the unit codes were painted out and Iraqi national markings and colours applied to the fuselage, wings and vertical tail surfaces. The fresh overpainting of the German identification in 65-colouring can be seen on the fuselage. The dark horizontal line was the dinghy release wire groove to the stowage in the tail cone. (M. Drewes via S. Santos)

RIGHT: This close-up view shows the extent of the marking, as well as the retention of the tactical white marking, which was used by the unit during September 1940 during the air battles over England. Officially, the marking should have been deleted, its use being rescinded in the last days of that same month. However, some of the unit's aircraft retained the marking in this reduced form when transferred east for the invasion of Greece. It remained in place when the unit then moved from there to Iraq in support of the uprising. (M. Drewes via S. Santos)

A Bf 109 E of the tiny Jagdgruppe Drontheim, which operated briefly from Norway; camouflage appears to be the final standard for the E-series, 71/02/65 with mottling on the pale side surfaces to break up their starkness. The visual clarity of the 'White 16' emphasised how effective were these air-to-air identification markings, even in the low light conditions seen here. (S. Santos)

Two photographs of a Bf 109 T-2 'Black 6', W.Nr. 7767 which was finished in similar style camouflage to that seen in the previous photograph, but with the Balkenkreuz on the fuselage applied only in white outline form. The very flamboyant addition of red to the spinner and along the fuselage, as far back as the cockpit, in the form of a long flame is unusual. Whether the flame marking was simply an embellishment of the red spinner colouring, as indicated by the spinner on the aircraft behind to the right, is not known. It might possibly have identified either the II. Gruppe or 2. Staffel since black had been used for the individual aircraft number. Note the small raised rectangular plate covering the oil cooling intake, on the underside of the cowling adjacent to the spinner's rear edge. The plate appears to be fitted with a small valve at the front edge. The aircraft behind was also fitted with the same device. As the oil tank was fitted above this area the cover plate may have had something to do with the fact that these aircraft were stationed in the Scandinavian region, where very low temperatures were encountered. Note also the 'C3' (100 octane) fuel coded octane triangle.

While the air fighting was in full swing over Britain in mid-1940, British forces in North Africa were fighting the first of several gruelling battles, initially against the Italians, later combined with the German forces. The first Luftwaffe aircraft did not arrive in North Africa until 30 January 1941, soon followed by reinforcements, all still in their European camouflage finish. While the aircraft of the permanent force were gradually re-camouflaged, subsequent transfers of units during the early part of the campaign continued to retain both their original camouflage and, as seen here, often their former tactical markings. This Bf 109 E (possibly 'White 2') was still marked with the yellow tactical colouring applied for its unit's role in the Balkans campaign. The very obvious wear on the propeller blades indicates that the aircraft had seen significant flying time. In the background, blurred but still visible, is the tail section of another Bf 109, its rudder painted in 04. The hasty temporary transfer of units during this period resulted in a mixture of anomalous tactical markings (see Volume One, page 151). (M. Pegg)

The transfer of permanent fighter forces to North Africa commenced early in 1940 with 1./JG 27 equiped with Bf 109 Es arriving on 18 April. Soon after, a desert camouflage of 78/79, with 80 mottling in some instances, was introduced. This was accompanied by a white band around the fuselage, used by Germany's Italian allies in the Mediterranean and North Africa (see Volume One, page 151). By the time these aircraft were found by Allied forces in a salvage collecting area near Gambut Main airfield, camouflage for all classes of Luftwaffe aircraft on permanent deployment to North Africa had been rationalised, with 79-colouring for upper and 78-colouring for lower surfaces. The fuselage white band remained standard for the Bf 109 Es, but a full suite of white markings had been introduced for their successors, the Bf 109 Fs and Gs. (J. Crow)

While the colour balance on this wartime photograph is too poor to be certain of the camouflage details, it does illustrate the visual effectiveness of the white band applied to the nose section (including also the spinner) of the Bf 109 Fs used in North Africa. The discolouration of the old yellow tactical marking below the engine cowling (the effects of long-term exposure to engine heat without repainting) produced an orange tone, seen here with a more reddish cast. This was another example of a previous tactical marking not being removed on transfer to another theatre of operations.

This better colour balanced photograph of a Bf 109 F, taken in North Africa illustrates the extent of the anomalous 04 yellow tactical marking beneath the engine section. The spinner has a yellow-coloured cover tied over it, but part of the actual spinner can still be seen – also marked in yellow – indicating that this was either a 3., 6. or 9. Staffel machine. Again, the yellow tactical marking, and its extent, indicates that this was another aircraft redeployed from a different theatre of operations. (D. Wadman)

During a head-on encounter, the effectiveness of the extended suite of white tactical markings (introduced for the North African campaign) is obvious from this almost perfect head-on shot of Lt. Mix's Bf 109 F-4 trop of 6./JG 27, which was shot down near El Alamein on 14 August 1942 (see Volume One, page 156). On the wing uppersurfaces, the spacing between the white sections of the Balkenkreuz shows that they were disproportionate, the centre section being oversize. Possibly the Balkenkreuz on the wing lower surfaces were also of the same incorrect proportions. (D. Vincent)

This closer view of the rear starboard side of 'Yellow 5' shows clearly the oversize central proportion of the Balkenkreuz, a feature seen on other Luftwaffe fighters, as detailed in Volume One. Camouflage colour was RLM 78 and 79. (D. Vincent)

LEFT AND OPPOSITE PAGE TOP: These two views of the front of Mix's aircraft show how carefully the white tactical marking colouring had been applied. The spinner had been removed for painting, as shown by the rear attachment plate of the spinner, which remained fixed behind the airscrew. Painting the spinner in situ would have introduced paint into the hub mechanism of the blades. The narrow panel fitted over the oil cooler had also been removed for painting, while the fixed section of the lower area had been masked out before spraying with white. The II./JG 27 Berlin Bear emblem had been applied using stencils, the first for the white shield shape, then a second for the black bear shape, then a final one for the red of the shield surround, bear's tongue and the 'crown' marking above the shield. Note the spaces left by the final stencil show the necessary 'bridges' between the cut-outs. (D. Vincent)

ABOVE AND OPPOSITE PAGE BOTTOM: These two views show a batch of Bf 109 F-2s captured during the retreat by German forces in North Africa near the end of the desert campaign. On the left is a mobile maintenance truck, marked with 'F.2' in white (registered WL-300443), which also has an unidentified unit marking on the front of both mudguards. In the background, the III. Gruppe Bf 109 F, marked 'I+4' in yellow, has standard 78/79 finish and a complete set of tactical markings in white, but the Balkenkreuz has a slightly oversize black centre. On the RAF trailer, 'Yellow 4' has an identical finish, but with a standard-size Balkenkreuz on the fuselage. Its markings identify that it belonged to a III. Gruppe, and there is evidence of painting out behind the number '4', possibly a Stammkennzeichen marking. However, the Bf 109 F in the distance, in identical camouflage finish, was marked with the alternative form of III. Gruppe marking, indicating two different Jagdstaffeln at the location. It, too, had the oversize black centre to its Balkenkreuz marking. The Go 242, in the shallow revetment on the horizon, helps to identify the location as possibly Tunisia, These large cargo gliders were used for the last phase of the operations in North Africa. Unfortunately, the canvas covers on the engines of all the Bf 109s make positive unit identification difficult but if the date suggested is correct, the aircraft probably belonged to III./JG 27 and III./JG 77. The only other III. Gruppe to have seen service in the fighting was III./JG 53, which had departed North Africa on 27 October 1942, with III./JG 77 replacing it.

Adoption of white tactical markings in the form of a white spinner, narrow band around the front section of the engine, fuselage band and wing tips (see Volume One, page 156) was a common feature of the F-model when aircraft were supplied direct from the factory wearing desert camouflage of 78 and 79-colouring. While the regulations specified a slightly narrower band width for the engine area, it was more convenient to use the panel line between the engine cowlings and the panel over the oil tank, as seen here on 'Yellow 5' of III. Gruppe. (S. Santos).

RIGHT: As the fighting became more intense in North Africa during 1942, more fighter units were transferred at short notice. In these instances, some Bf 109s arrived in full European camouflage, as seen here on 'Yellow 16' of III./JG 77 which had begun to arrive on 28 October. The presence of just a single white band around the fuselage reflects the unit's previous deployment on Crete. The size of the '16' was unusual, as other aircraft from this unit wore standard size numerals. Equally, the location of the III. Gruppe emblem on the fuselage behind the cockpit on the starboard side, was not common. The yellow-coloured wording above the emblem is not clear but below appears "Ubben", a reference to the III./JG 77 unit commander, Hptm. Kurt Ubben. Note that both the numerical marking and the Gruppe emblem were applied over an area where a previous marking had been painted out. The JG 77 emblem of a red heart on a white diamond with dark surround can be seen on the engine cowling.

LEFT: This frontal view of the same aircraft, flown by Uffz. Lindemann, shows the III. Gruppe emblem in better detail. The dark edged surround to the red heart appears to be red rather than yellow, in the Gruppe emblem of II. Gruppe/JG 77. Possibly Lindemann had been transferred from the latter to III. Gruppe but retained his connection with his former commander via the wolf's head emblem and wording. Note the lack of any tactical markings on the front section of the aircraft other than for the use of a spiral spinner, in this instance in 70 judging by the colour comparison with the propeller blades. In 1942, spiral markings were still optional at unit discretion rather than having any official connection with tactical markings, something that changed in June 1944 (see Volume Two, page 293).

This view of very early pre-production Fw 190 A-0 and A-1 aircraft at the parent plant shows the variation in camouflage. The A-1 airframes were finished in the revised camouflage of 74/75/76. Note the Fw 191 bomber prototype in the background. As with the two A-0 series in the left foreground, the overall finish was 02. The apparent segment of dark colour is simply the effect produced by different panel sections. The two A-0 machines had their undersurfaces and entire engine cowlings camouflaged in 65.

The single-engine fighter force was augmented in 1941 when the Fw 190 entered production, the Fw 190 A series being manufactured at a number of sites. This production line photograph at the Bremen factory illustrates the final assembly stage, where camouflage was not applied until after completion of the entire airframe. The overall colouring at this stage was a mixture of 02 and 76, with some wing components still in 02 (the filler paste on the panel lines around the leading edge of the wings can be seen clearly), while the engine cowling was finished in 76 (compare the slightly darker shade of the 02 on the inside of the open panels). The 02-colouring of the undercarriage legs and the inside of the oleo doors also provides a colour reference. The wheeled cart visible below the first aircraft has 'Serien-bau 189' marked on it in white, probably identifying the production line facility. Serien-bau literally translates as series production.

Production of the Fw 190 A-4 series at the Bremen plant received major sub-contracted components pre-painted in camouflage, ready to be married to the fuselage structures. Here, the tail empennage (another sub-contracted assembly) is being attached to the main fuselage structure, which was produced by the parent firm. Note the variation in uppersurface camouflage colours on the fuselages. The variation in the freehand-applied mottling in 74 and 75-colouring on fin and rudder is also well illustrated. In contrast, the precise positioning of the Hakenkreuz and Balkenkreuz markings (painted with the aid of stencils set at defined points on the structures) is clearly seen.

This overall view of the wing assembly line on the right affords the opportunity to see how freehand-applied 75-coloured segments varied slightly between each set of wings. Again, the precise placement of the wing Balkenkreuz is obvious.

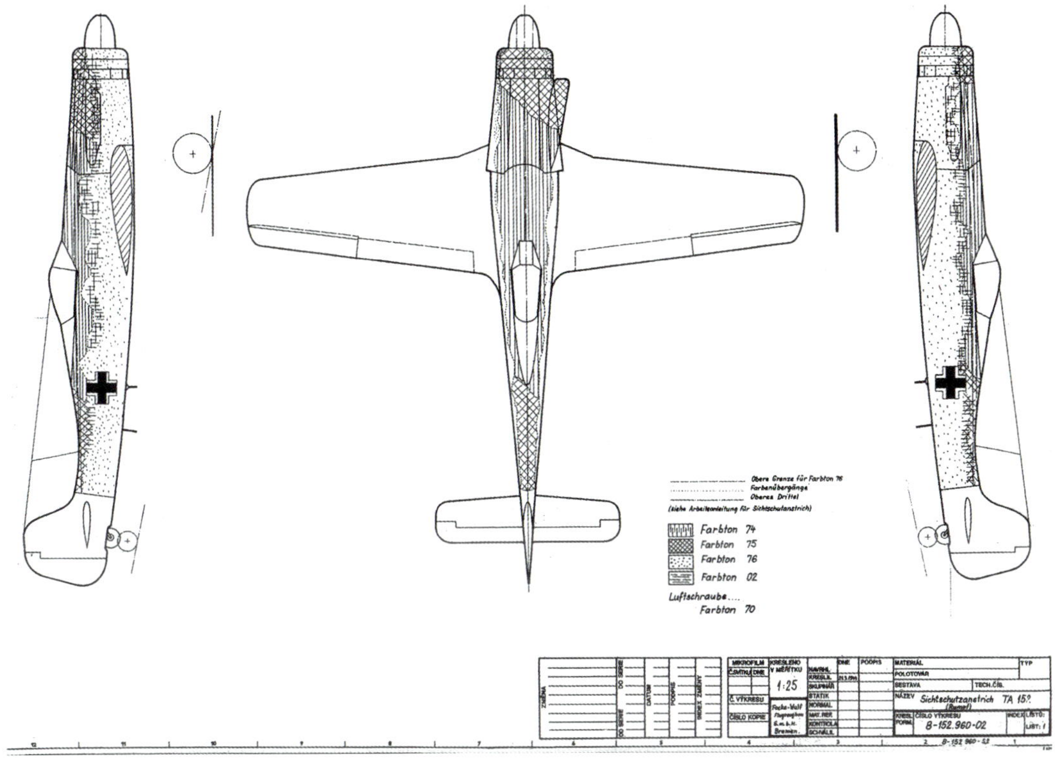

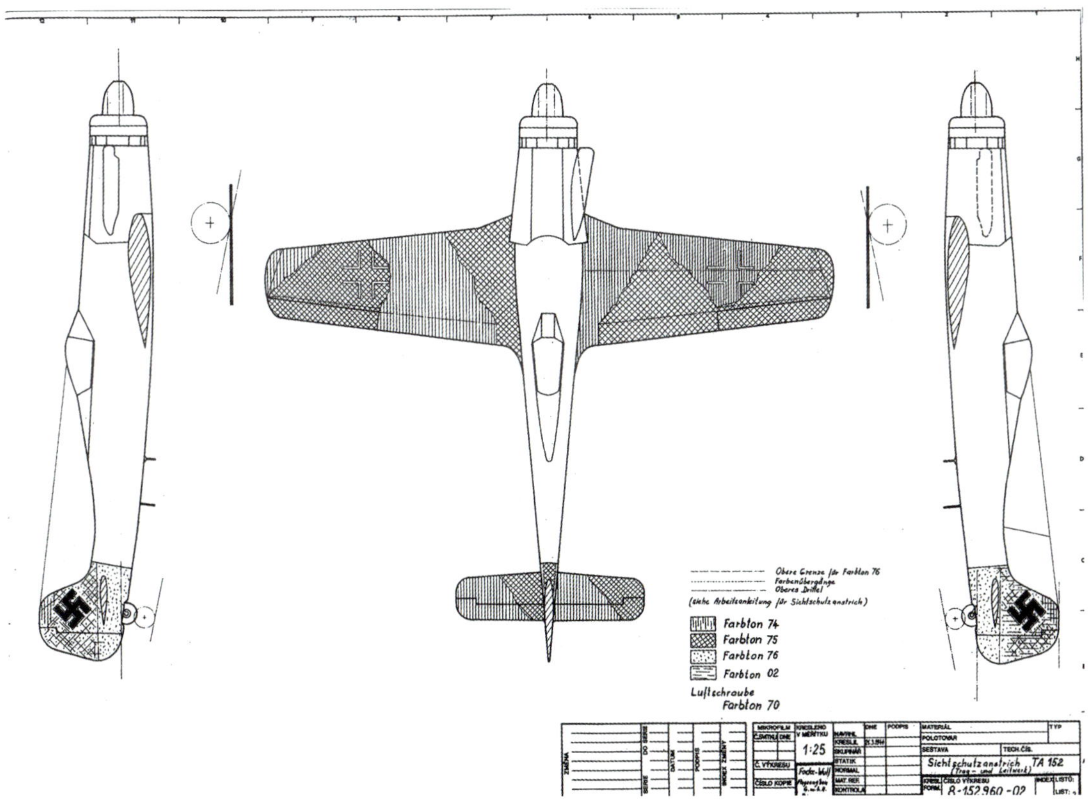

The process of sub-contracting had been put in place from the very earliest days of the National Socialist government's reign in Germany. Right from the earliest period, the idea was to decentralise the production process so that if and when bombing commenced, disruption would be minimal. Thus, the process had been established within the aircraft industry of the parent company manufacturing the main structure with sub-contractors supplying everything from the smallest part up to the main additional components such as wings and empennage structures. Each main manufacturer in the aircraft industry had its own processes of assembly and painting, as noted opposite and detailed in Volume One. Issue of the relevant camouflage diagrams for each type was controlled by E-Stelle Travemünde and, as far as is known, only camouflage diagrams of complete aircraft were issued, at least until well into the war period. The only known example of variation so far located to this long-established practice relates to production of the Ta 152. A complete camouflage diagram was issued as normal (see Volume One, page 183), and this was modified as changes to colours and coverage were revised by the RLM. These two diagrams are thus unique in showing that two separate drawings also were issued – one to the parent company detailing the camouflage for the main structure; and one to the sub-contractor who produced wings and tail assembly showing the camouflage for those components. This new system was significant because it allowed pre-painting of major components at disparate locations before all parts were brought together at the main parent manufacturing plant. This tied the sub-contracted companies into the main final camouflage process but, by default, in the very late war it became the cause of the sometimes mismatched camouflaged components seen on aircraft leaving the main assembly plant. Changes in camouflage colours sometimes outpaced changes to processing, and the delivery of camouflage paint stocks to different sub-manufacturing plants did not always keep pace with that of the main assembly plants. (JaPo)

Here we return in time to events in the run up to the invasion of Russia in mid-1941. Although colour balance is poor, this photograph nevertheless yields some useful details. The colour shift has produced a blue-grey effect, but the actual colours were 71/02, with a rather unique form of sharp-edged segments of 71 to break up the 65-colouring of the side surface colouring. Note the Schlacht black triangle marking on the aircraft at the rear, which also has a soft-sprayed darker application on its side surfaces. 'Black 10' in the foreground has a line of at least seven kill markings painted along the upper section of its rudder. As part of the force used for Operation Marita in April 1941, the aircraft wear the full set of allocated tactical markings – 04 yellow on engine cowling, wing tips and rudder. Unit emblems can just be discerned on the cowling of 'Black 10' (see Volume Two, page 274).

RIGHT: This closer view of the engine cowling of another aircraft from the same unit reveals that two emblems were carried – the Wolf's Head marking of III./JG 77; and a bird emblem not dissimilar to that used by II./JG 54 (possibly a connection with the pilot's previous unit). The red and white-segmented spinner cap would appear to identify 8. Staffel.

By 1941, the Luftwaffe had settled into a static war along the western seaboard facing Britain, and tactical markings for single-seat fighter aircraft were quite bold, as seen on this Bf 109 F. The initial practice was to retain the 04 yellow for the entire engine cowling for the front of the aircraft, a carry-over from the last stages of the air fighting over Britain the previous year. In time this would be reduced to yellow on the ventral engine panel area. The lacquer used for these tactical markings was by then a permanent type as seen here, with a quite smooth finish. The red spinner was most probably the 2. Staffel colour because the emblem, a black hand grasping a Spitfire, appears to be a simplified version of the I./JG 52 Wappen. In its full form, the hand and aircraft were superimposed on a shield marked with a map of England and the western seaboard of Europe. Note the white alignment mark across the spinner back plate (still in 70-colouring) and spinner joint line. The propeller blades show evidence of considerable wear.

The first winter of the Russian campaign severely tested both men and equipment. This line-up of Bf 109 F-4s of II./JG 54 have had their oleo leg coverings removed to stop snow clogging them and causing wheel retraction and lowering failures. The second machine in line has a red spinner, indicating that it belonged to 5. Staffel. All the aircraft, including the Ju 52 in the background, have temporary white winter applied, although on 'White 3' it appears to have less solid, creating a soft mottle effect.

Not all of JG 54's Bf 109 F-4s were given winter camouflage immediately. Photographed at Siwerskaja in the winter of 1941/42, Major Hannes Trautloft's aircraft, marked with its Kommodore's chevron and bar, carried an incomplete suite of tactical yellow markings. The rudder retained its camouflage colouring and the 26 white kill markings. The very poor colour balance does not allow any definitive remarks about the apparently peculiar camouflage, but most probably it was simply 71/02. Some Bf 109 Fs retained the old colours for their deployment to other locations where the mixture of greens better served their purpose. JG 54 was another select unit associated with a range of experimental camouflage schemes (see Volume Two, page 278).

Bf 109 Fs of II./JG 51 had just arrived at their new base at Orel, on the Eastern Front, when a surprise attack by IL 2 and Pe 2 aircraft caught them without cover, causing considerable damage. The aircraft had just transferred from Norway and appear to have retained just the yellow band marking of that Front. No other yellow markings (beneath the engine and under the wing tips) can be seen. The camouflage on the aircraft in the right background appears to be quite dark, but the apparently 'white' rudder was an illusion produced by the strong sunlight on the angled rudder. (S. Santos)

This side view of the same aircraft (above), shows a camouflage of 74/75/76 with heavy mottling to the side surfaces, the two greys almost obscuring the 76-colouring of the rudder. (S. Santos)

'Black 9' of 4. Staffel (just visible in the background of the previous photograph) shows the heavy mottling used on these machines, especially on the fin and rudder where 75-colouring had eliminated any trace of the 76-colouring, other than around the Hakenkreuz, before addition of the 75-coloured mottles. This heavier application of mottling probably accounts for the dark appearance of the machine seen in the background of the first photograph. The yellow band marking is seen clearly here. (S. Santos)

JG 54 had a long-standing link with camouflage development field trials throughout much of the war. This Fw 190 illustrates the ongoing experiments using the pre-war colours of 61, 62 and 64, all of which would reappear in 1944 as mainstream camouflage colours under the revised designations of 81, 82 and 83 (see Volume One, page 107). This machine, 'Black 8' photographed in early 1943, has what appears to be a combination of either 61 and 62, or 64 and 62, on its uppersurfaces and fuselage sides, colours that were much more useful within the vast Russian landscape than the contemporary greys of 74 and 75. Lower surface colouring remained standard 76. The theatre of operations' tactical colouring was yellow used under the engine area, wing tips, around the fuselage as a band and on the rudder. However, by this stage of the war in Russia, the yellow band had been repositioned directly behind the fuselage Balkenkreuz, thus providing identification from above and below without compromising too much the primarily ground defensive colouring of the camouflage. The full rudder colouring also had been reduced to just the section visible from below the horizontal tail-plane for the same reasons. Note that, although snow still covered much of the countryside visible in the distance, no temporary white finish was in use on this particular machine, indicating that it may have been only recently delivered from Germany. The almost pristine condition of the propeller blades supports this possibility (JG 54 had only begun to re-equip with Fw 190s in December 1942). In direct contrast, the Fi 156 marked SB+VG in the background was finished overall in temporary white winter camouflage.

This view of the front portion of a Bf 110 F illustrates the density of the mottling used when camouflage requirements had shifted to the standard day fighter scheme colouring of 74/75/76. Note the use of mottling also along the leading edges of the wings to break up the pale 76 colouring where the undersurface colour had been taken around the leading edges – a practice started in 1940 to break up the dark line of the wing when seen head-on. The mottling has also been taken right down the sides of the engine cowlings as far as the sides of the radiator cowlings. There is no evidence of any tactical marking colour beneath the wing tip, ruling out the Eastern Front as the location. Spinners were also in RLM 70, but the one visible had a miniature of the standard requiring spinners to be marked one-third white/two-thirds black – an interesting but unexplained item.

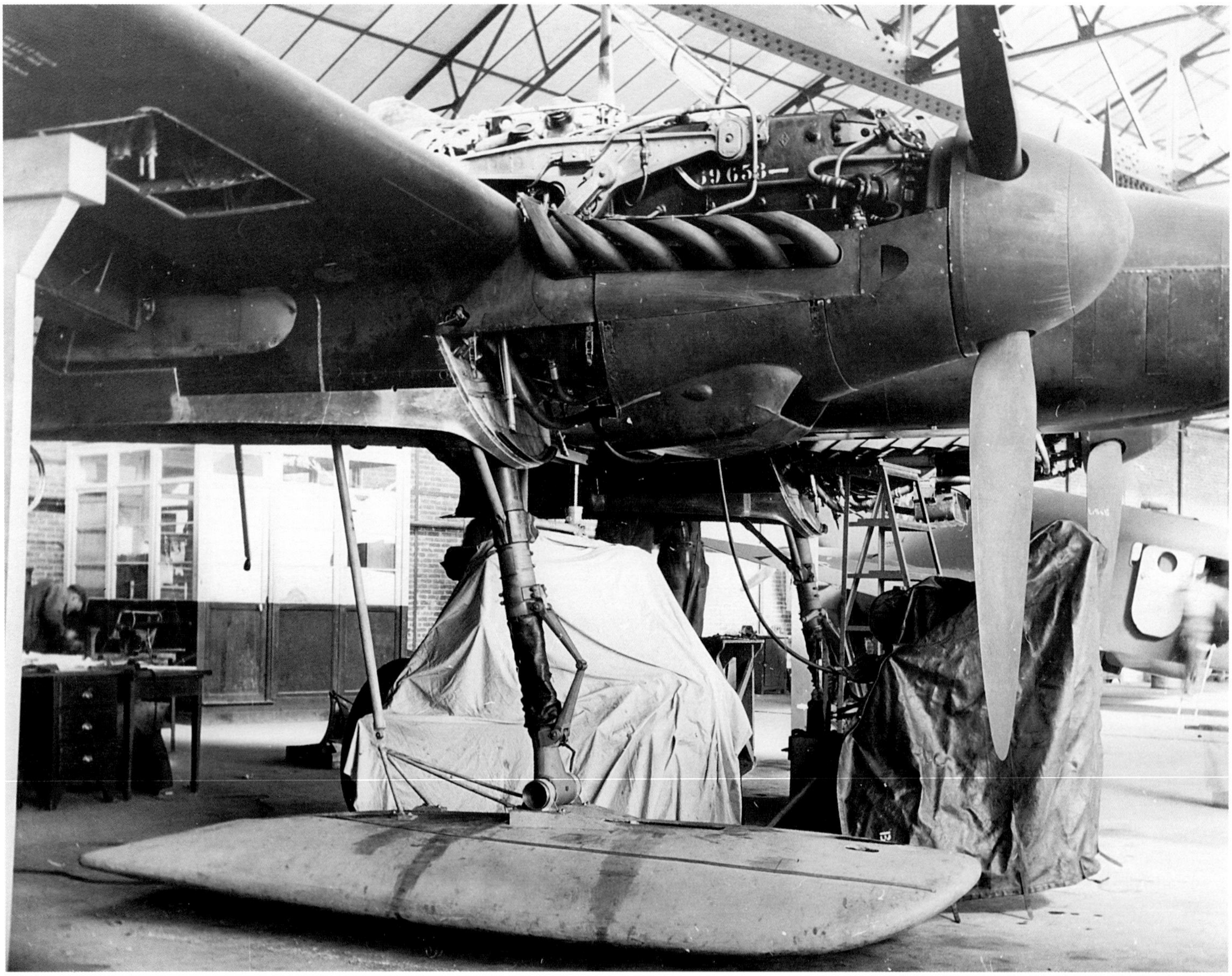

ABOVE AND OPPOSITE PAGE TOP: These two photographs illustrate an interesting experiment aimed at improving Bf 110 snow take-off and landing capabilities, logically with the extremes of the Eastern Front winters as the focus. The savagery of the Russian winters had caused problems for several types of Luftwaffe aircraft. Those fitted with fixed undercarriage and spats rapidly discarded the latter to prevent the wheels from seizing up through accumulation of ice and snow. For larger types such as the Bf 110, the problem of wheels freezing and seizing on their axle while retracted was a problem which required a better engineering solution. The skis appear to be very similar in design to the type used for the Siebel Sh 104.The colouring on this Bf 110 F is interesting. Judging by the RLM 02 colour match with specific parts, i.e. the undercarriage oleos, ski assembly, wing radiator core and engine bearers, the dark overall colour of the main airframe was not in that primer colour. The spinner and propeller blades (one on the port engine hand painted in white with what appears to be E.16.35), both in regulation RLM 70, also rule out that colour for the main airframe, although they appear to match the removable upper cowling over the gun bay in the nose. The original primer 02-colour can be seen where the fixed side panel of the engine cowling had been removed. The overall colouring appears to closely match the tonal value of the engine block (marked '69653' in white), a dark colour similar to RLM 66, while the undersurfaces of the wing were marked with neat white stencilled maintenance instructions, indicating that the overall paint finish was permanent. The small portion of the Balkenkreuz provides a tonal measure for pure black, and appears slightly darker than the background colouring of the airframe. This was most probably a result of the type of lacquer used for national markings, an enamel that had a slight sheen to its surface whereas the camouflage black, though smooth, was far less reflective. The last two letters of the Stammkennzeichen or code group –WQ – appear to be in RLM 77. This combination of black overall plus 77-coloured markings points to the possibility that this was a Bf 110 F-4 night fighter airframe, although its choice for the modification may have been irrelevant to the experimental nature of the work. On the Eastern Front, lack of ground radar facilities made night fighting sorties fairly irrelevant to the main battle. That, in turn, may have influenced the choice of this particular aircraft. The Caudron C445 A-1 in the background had standard 70/71/65 camouflage with an 04-coloured Eastern Front yellow band, confirming the location of the workshop hangar. The Stammkennzeichen most probably was DL+XM, in which case the Werknummer would have been 1048 – part of a batch of Caudron C 445s produced between February and April 1943 (DL+XM to +XQ with Werknummern 1041 to 1052). This provides a base time reference to the photographs, placing the activity in the latter part of 1943. The aircraft had its French-type name, 'Goerland', marked in black beneath the cockpit windscreen front line, and the engines had a black cheat line extending from spinner to wing root. The staff visible wore no heavy clothing and the hangar workshop appears to be open to the outside, judging by the bright sunlight from the right; all of which indicates the mid-1943 period.

The fighting in North Africa had ebbed and flowed for much of that difficult campaign, and damaged aircraft often had to be abandoned in the haste of retreats. This Bf 109 F was undergoing maintenance when abandoned (note the replacement starboard horizontal tailplane in the foreground), with its tail unit damaged and its engine under repair. The double chevron identifies it as the mount of a Gruppenkommodore and the absence of any marking aft of the Balkenkreuz identifies it as I. Gruppe, either of JG 27 or JG 77. Only the fuselage white band is visible, but it should have carried the normal full range of white spinner, nose band, wing tips and fuselage band. Arrival of the F variant coincided with the revised suite of tactical markings introduced for that theatre of operations. Note that the fin had been oversprayed with 79-colouring, leaving a pale area behind the Hakenkreuz, indicating that this was one of the early deliveries of the type that had been repainted, in desert camouflage, at a Luftwaffe maintenance depot before delivery to its front line unit. Factory painted Bf 109 Fs allocated to North Africa were finished in 78 and 79, complete with tactical markings once allocations for that battle front had been established, while initial deliveries had to be repainted post-production. The style of application of the two colour division also points to repainting, production deliveries having either a division along the horizontal mid-line, or the uppersurface colour taken right down to the lowest point of the fuselage sides. (S. Coates)

ABOVE: The last stages of the battle in North Africa ended at the Tunisian bridgehead as the Luftwaffe and ground forces sought to extricate themselves to the other side of the Mediterranean. This abandoned collection of badly damaged aircraft, probably assembled for salvage at El Aouina, offers interesting detail. In the foreground are three Bf 109 F airframes – all apparently I. Gruppe machines – consisting of 'Black 8', 'White 1' (with a patch behind the digit where a numeral had been painted out), and one marked with the Gruppe TO (Technical Officer) symbol. Camouflage on all three was 78 and 79. Behind is an Fw 190, marked with a Gruppenkommodore's symbol and still in European camouflage of 74, 75 and 76, but with white applied to the engine ring, indicating that it possibly also had a white fuselage band marking. The late arrival and relatively brief service of Fw 190s into this battlefront saw some with only partly marked tactical markings. To the right is an Hs 129 B-2 trop of 8.(Pz)/Sch.G 2 (transferred under its original designation of 5./Sch.G 1), the unit having been hastily transferred from Jessau in East Prussia. Camouflage for these aircraft was applied at a Luftwaffe maintenance depot using a variety of styles, but all employing the simple expedient of an application of 79-colouring over the existing 70/71 camouflage. (S. Coates)

BELOW: JG 27, after long and hard fighting in North Africa, was redeployed to the Reichsverteidigung (the defence of the Reich force) in late 1943. This Bf 109 G-5/R-6, 'White 9' of 7./JG 27, had an all-white empennage marking, just visible at the edge of the photograph, and a narrow white band. The latter was one of two colours in use by theReichsverteidigung during 1943 (see Volume Two, page 288), Bf 109s of 7. and 9./JG 3 and the Me 410s of ZG 26 also using this form and colour of marking. Camouflage was the standard 74/75/76 of the period.

BELOW: The same aircraft (left) is seen here in formation with 'White 2' and 'White 7', escorting an He 111. The nearest two aircraft have the reduced width band marking of the Reichsverteidigung force, while 'White 7' retains its North African tactical marking suite of a broad white fuselage band with a narrow white band around the front section of the engine cowling (and presumably white wing tips). 'White 9' and '7' both wear the III./JG 27 unit emblem on the engine cowling as well as the 7. Staffel emblem marked below the cockpit. The visual effectiveness of the all-white empennage marking is obvious; 'White 2' has a white rudder, another example of this specialist form of marking, providing a rare combination of two aircraft from the same unit carrying such markings (see Volume Two pages 285-287).

These two photographs show Bf 109 G-6s, belonging to I./JG 27, on the airfield at Fels am Wagram in Austria around August 1944. All wear standard camouflage of 74/75/76 and, with the exception of 'White 10' seen in the lower photograph, they have the revised Balkenkreuz marking composed of white angles with the darkest of the two camouflage colours brought down to fill in the centre of the marking, as per the official directive issued, originally mid-war and reiterated in the 1944 edition of Der Flugzeugmaler (see Volume Two, page 441). The black centre and narrow black outline-form of the Balkenkreuz marking seen on 'White 10' indicates that this was an older aircraft. The wide band marking was green, part of the initial Reichsverteidigung system used by aircraft of Jafü Hungary, part of Luftflotte 4 which also controlled Jafü Ostmark established in 1944 to cover the southern defences of the Reich with fighter aircraft positioned in southern Austria and over the border in Hungary (see Volume Two, page 288). Rudder colouring was 04 yellow, standard for the 'eastern' defence forces of the period. Werknummer 140255, nearest the camera, provides detail of some of the range of markings seen on Bf 109s of the period. The fuselage frame numbers, always marked on the port side, were marked in black. The jacking point marking, however, had disappeared beneath the wide green band marking; the latter probably being 90 cm wide as per the instructions for this type of marking. The small oval fuel tank access panel appears to be a replacement part, still on 02-colouring. All the aircraft had been fitted with Rüstsätzen R3 and R6; MG 151/20 mm cannon in gondolas beneath the wings (R6), and a belly rack to carry a 300 litre drop tank (R3).

Displaying the very minimal of tactical markings – a spiral on the spinner – identifies this as one of the fighter force on the Western Front after the official introduction in June/July of the spinner spiral marking for all fighters engaged on the Western Front (see Volume Two, pages 293-294). Camouflage was the greys of 74/75/76 and close examination of the paintwork shows that it had been in service for a considerable time. The Balkenkreuz on the fuselage had had its white angles toned down with a dark colour, probably 74 in this instance, applied as a soft-sprayed application. Unfortunately the unit identity is unknown and just the bottom tip of what appears to be a white-outlined '7' is all that is visible on this I. Gruppe machine. Note the lines of camouflage netting on either side of the aircraft, some coloured green, the other a brown shade, colours that would soon be introduced for the camouflage of Luftwaffe aircraft in general. (Royal Danish Air Force via M. Jessen)

I./JG 1, as part of the Reichsverteidigung, adopted the generic wide red fuselage band marking introduced in the Autumn of 1943 for that force. (By December 1944, most Jagdgeschwadern involved in the Reich defence force had been allocated individual unit band markings and colours - (see Luftwaffe Colours, Jagdwaffe Volume Five, Section 3, Defending the Reich, 1944-1945, pages 220-221 for full colour details - but this aircraft appears to be from the period prior to that final move.) This JG 1 Fw 190 A-7/R2, 'Black 10' outlined in white, was fitted with an ETC 501 rack beneath the fuselage, capable of carrying a 300 litre drop tank or bombs. Camouflage appears to be 74/75/76 rather than the later camouflage colours, something reinforced by the style and proportions of the fuselage Balkenkreuz marking, which was of the pre-mid-1943 style, retaining a black centre and black edging to the white angles. The area behind the '10' marking appears to have been painted out, most probably to obscure an earlier marking – again pointing to an older aircraft. The colour used for the painting out looks like RLM 74. The small patches of primer red on the fabric of the rudder and the port aileron were repairs to what may have been bullet damage, given the size and distribution of them. The rudder was yellow – again emphasising the 1943-early 1944 period before the Reichsverteidigung system moved to its logical conclusion in late 1944. The wing tips appear to be also marked in yellow, the starboard tip being more clearly seen in the strong sunlight conditions. It is possible that this aircraft was W.Nr. 430465 of 2. Staffel, which was shot down on 20 February 1944 over Eschwege, near Kassel, killing Hptm. Harald Römer.

This Fw 190 A-8/R2 of JG 300 has an interesting combination of dark, ground defensive camouflage combined with an oversize Hakenkreuz. Such anomalies with the size, but not proportions, of Hakenkreuz occurred on both fighter and some other classes of aircraft, such as Ju 88 night fighters as well as some day bombers, notably He 177 A5/R6s of II./KG 40, equipped with Hs 293 A wire-guided bombs for anti-shipping operations from bases in France. The camouflage of 'White 3' (its tone matches that of the slightly subdued white of the fuselage Balkenkreuz) appears to be a combination of 82 and 83 with some random mottling in a darker colour, logically 81. Note that the dark camouflage combination was taken right down the fuselage sides to the lowest extent, even to the point of adding mottling to the blisters on each side of the engine cowling. It also overlaps the Hakenkreuz in parts, the original pale background being apparent. The empennage unit may have been delivered in the older style of plain 76-coloured background and then oversprayed to match the main fuselage structure. There appears also to be a white disc marking aft of the Balkenkreuz, the old IV. Gruppe-style of identification, dating back to the pre-war period. If this is indeed so, then it was at odds with the contemporary flattened 'sine wave' form of marking adopted for that formation. The absence of any form of Reichs defence marking dates the photograph to around late 1944. The weather appears to support a late Autumn-type of day. (J. Crow)

An interesting if enigmatic marking; this Bf 109 G had seen considerable air time judging by the extensive wear on the rear face of the propeller blade that is visible, before being captured at the end of the war. Camouflage is fairly standard in application; 81/82 with mottling in both colours over the 76-coloured side surfaces, although the side surfaces aft of the cockpit appear to have been deliberately toned down with a wash of slightly darker colour that matches the base colouring of the rudder. The fin has a base colour of 82, with a solid segment of 81 extending upwards, at an angle, from the rudder mass balance top line. The rudder has large mottles of 81, with a single one of 82 over the base colouring, which appears to be a thin coat of 02 over the original 76-colouring, leaving a wedge marking at the tip. The latter appears to be either white or yellow, its shape, size and position strongly reminiscent of the first style of yellow tactical marking introduced in August 1940, during the air fighting over Britain (see Volume Two, page 266). There seems no other logical reason for such a distinctive marking appearing on 'White 12' other than as some form of air-to-air recognition device. Its use was not constrained to just this aircraft as the photographs on pages 38 and 39 illustrate. The radio access hatch had been replaced with one from another aircraft. The Balkenkreuz and Hakenkreuz were of the pre-July 1944 style, indicating that possibly this was a machine built in the latter half of 1944 when the change in markings was still slow to be adopted. (J. Crow)

Surrendered at Fürth at the war's end, this Bf 109 G-14, 'Black 17', W.Nr. 463141, appears to have a yellow-coloured rudder, but this was just the effect of the strong sunlight at a low angle as the rudder was held over to counter the torque from the engine during the landing run. Camouflage was 81/82/76 with mottles in both colours. Not easily seen is the white wedge shape marked on the tip of the fin, identical in shape and position to the one seen in the photograph at the top of page 37, demonstrating that this marking was used by more than one aircraft (though possibly both may have belonged to the same unit). The lack of tactical markings other than for the spiral spinner denoted its deployment had been confined to the Western Front. (J. Crow)

RIGHT: This composite colour photograph, of 'Black 17', though poorly colour balanced, is good enough to determine that the camouflage was 81/82/76. Note the now clearly visible wedge marking at the tip of the fin, which is definitely white (compare with the 04-coloured fuel triangle), as well as the fact that the rudder, far from being a pale colour, was a solid 81/82 colouring – a further reminder that apparent colouring seen in black and white photographs can be very deceptive at times. (J. Crow)

BELOW: This closer view of the starboard side of 'Black 17', shows that the white segment was definitely present on both sides of the rudder, confirming that it had some function as a distinctive marking. (J. Crow)

ABOVE AND BELOW: Two views of Bf 109 G-14AS, 'White 4', W.Nr. 6316, after its surrender at Kassel in 1945, reveal an interesting example of the mismatch of major component camouflage colours. The main airframe camouflage appears to be in the older 74/75/76 scheme, but the entire empennage was painted in 81 and 82. The 82 had been used as the base colour with spots of 81 added over it. The rudder had had short strokes of 76-colouring added to break up the dark mass. Some attempt had been made to integrate the two camouflage schemes by extending some of the 81-colouring onto the uppersurface of the fuselage, immediately adjacent to the empennage. The Hakenkreuz on the fin was the correct late war, white outline form, while the Balkenkreuz on the fuselage was the old mid-war type with a dark centre in 74-colouring. This supports the contention that camouflage was 74/75. The Werknummer had been stencilled on each side, in black, using just the last four digits of the complete number. (J. Crow)

Photographed near Magdeburg in May 1945, this well-balanced colour photograph reveals an interesting array of colours. In the foreground is 'Yellow 8', an Fw 190 of 6./JG 301, marked with the unit's distinctive yellow and red band combination. Camouflage appears to be 81/83 rather than 81/82, judging by the dark tone of the green colour on the open cowling in front of the cockpit, as well as on the fin, which looks too dark in tone for 82. (Enlargement of the mottle on the fin shows a distinct olive cast to the edges, further indicating 83 rather than 82). The 81-colouring can best be seen directly under the cockpit quarter-light area and above the number '8'. Compare the pale background colouring of the fin, behind the mottling, with the bright blue of the rudder, the 76-colouring of the fuselage having been toned down with an overspray of 02. Note also the pale grey-white 76-colouring of the undercarriage oleo doors. The rudder fabric has been vandalised with sections, probably with kill markings, cut out, indicating that the pilot had a reasonable score. The addition of dark mottles of 83 on the engine cowling had been done with a soft spray of closely grouped mottles to form an almost solid appearance. Again the paleness of the 76 grey-white colouring can be seen in small patches, no doubt the reason for the heavy mottle finish to the area. The oddity in the unit markings is the use of yellow for the aircraft's number, but white for the II. Gruppe horizontal bar. Normally it would have been displayed as shown on the Bf 109 G immediately behind, another 'Yellow 8' of a II. Gruppe, this time JG 27, identified by its broad band of RLM 25 green. This machine appears to have a camouflage of 81/82 and retains a black spinner without the spiral. The fin and fuselage aft of the band marking appear to have been oversprayed with 02. The dark area over the number '8' is a fine netting, the sunlight shining through holes in places. Unlike the Fw 190, the Bf 109 G retains the older style of Balkenkreuz and Hakenkreuz, possibly an indication of its relative age.
To the left are the remains of two Go 242 fuselages (??+AW and ??+BI), finished in 70/71/65, each marked with a yellow tactical band.
(Walsh via J. Crow)

For the Luftwaffe in 1945, the deadliest peril was on the ground where there was no escape from constant ground-attacks by Allied aircraft. At field dispersals, some semblance of cover could be obtained by using netting over parts of the aircraft (especially the national markings), tree branches or simply dispersal behind the edge of tree-lines. On the end of the concrete runway at Brandis, east of Leipzig, little could be offered in the way of disguise. However, the diminutive size of the Me 163 B allowed almost the entire aircraft to be hastily covered with canvas tarpaulins, not just as weather protection but also as an aid to camouflage. The few areas left exposed show only heavy mottling in 81 and 82 over a 76-coloured background. The tip of the nose cone and spinner were yellow.

'Black X', W.Nr. 500200, of 2./KG 51 photographed at Fassberg in May 1945 after its surrender, along with other Me 262s. Currently held by the Australian War Memorial, its original colours are still preserved: bare metal surfaces for all but the nose cone (sheet steel, which had an undercoat of 02), other than for control surfaces supplied by a sub-manufacturer; the rest of the airframe had just two coats of paint applied directly over the translucent metal etch finish, 81 first of all with segments of 82 added last, thus darkening the tone of the 82 green. The effects of lighting levels and angles produce marked optical effects when assessing colours, and this particular aircraft offers numerous examples to illustrate this phenomenon (see Volume One, pages 19-20).

Bf 109 K-4, 'White 8', from an unidentified 1. Staffel, which, in company with a Bf 109 G-10 (or G-14), made a crash landing behind Allied lines at the end of the war (see Volume One, page 184). Camouflage was quite dark, indicating a combination of 81/83 rather than 81/82. Using magnification and colour shifts it is possible to pick up a linear loop of the darker colour at the rear of the fuselage, as well as several distinct mottles of the same colour around the area just above and behind the wing root, but the application was such an intense mix of the two colours sprayed over each other that the camouflage pattern on the side surfaces is difficult to establish. The Balkenkreuz on the fuselage is a white outline form, with a wide panel of 83-colouring used as a background to fill in the centre. This can be seen extending below the horizontal arms of the marking, and down over the paler undersurface colouring. The fin and rudder unit in particular were quite heavily covered with a mixture of the two colours, one large darker mottle being faintly discernible on the rudder, in line with the horizontal tail surfaces. The Hakenkreuz is not easily seen at first due to the dark dense camouflage on the fin. The belly-landing was made in quite yellowish clay (visible in the foreground), which sprayed back over the undersides as a thin yellowish wash in places, making the 76-colouring look quite different in colour. (J. Crow)

A multiple example of the varied camouflage schemes in use at war's end. Found at Braunschweig, in the left foreground is 'Red 10', a Fw 190 A-8 of an unidentified I. Gruppe, camouflaged in 81 and 82 on the extreme uppersurfaces, with the 76-coloured side surfaces (and cowling section beneath the engine) over-sprayed with a light coating of 02, a common feature in the late war period to tone down the stark pale colouring to provide better ground camouflage. Mottling was both minimal and very soft on the fuselage side surfaces, but more intense in colour on the empennage, a sub-contracted unit delivered already camouflaged. Note the solid line of 81-colouring down the leading edge of the fin. The undercarriage oleo covers were bare metal, confirming that the undersurfaces of the wings were also bare metal for most of their area, a revision introduced onto the type in the previous October to conserve materials and labour. The style of Balkenkreuz and Hakenkreuz markings point also to this being a late production airframe, with a solid black Hakenkreuz specified for pale surfaces (which the fin was, with minimal mottling in 81 and 82) and an outline form Balkenkreuz. Other than for the spiral spinner marking of white on black, no other form of tactical markings were carried, indicating that this aircraft had been used on the Western Front (Reichsverteidigung markings were the only extra used on that Front but by 1945 many units seem to have become less inclined to mark them, possibly because of the pace of battle – see Volume two, page 290). Immediately behind the Fw 190 is a Ju 188 D-2 camouflaged in 71/72/65 (production of the D-type ended before stocks of the old colours were exhausted – see Volume One, page 115), with yellow tactical markings below the wing tips. The code cannot be discerned but the individual aircraft letter appears to be 'Y' marked in RLM 25 green and thinly outlined in white, possibly indicating a Gruppe Stab machine though the use of the letter 'Y' for Stab marking remains enigmatic – unless this was an Erg. Gruppe). The spinner tips are also painted RLM 25 back to the line of the propeller hubs. On the extreme right of the photograph is a Ju 88 G-4 night fighter, also marked with yellow beneath its wing tips. The overall base colouring was 76 with 75 on the extreme uppersurface areas, over which a network-style of predominantly 82 green strokes had been applied to the upper and side surfaces with small strokes of 81 added. The code is not discernible in full, but the last two letters were 'FB', marked in black, the aircraft's Stab status being reflected only by the RLM 25 green applied to the spinner tips. The Balkenkreuz below the wings had been applied as a solid black cross, without white edging, a feature not uncommon on single-engine fighters in the late months of the war, but not usually seen on twin-engine machines.

Night Fighters

A rare colour photograph of a Bf 110 F, marked with the prominent emblem of I./NJG 1. The spinner tips were in the II. Gruppe colour. The slight sheen on the surface of the nitro-cellulose lacquer can be seen. This permanent finish, with its smooth surface, added several kilometres per hour to top speed. The angled side-frame of the front windshield was missing when the photograph was taken.

Bf 110 E, G9+GA, W.Nr. 392(5?), of the Stab NJG 1, photographed at Düsseldorf in winter conditions. The black overall finish was the standard nitro-cellulose lacquer with code letters in RLM 77. A line of kill markings was painted on the fin in what appears to be white or yellow, partly covering the Werknummer. The presence of the yellow band around the rear of the fuselage is unexplained. (W. Falk via M. Jessen)

Not the same aircraft as on page 43 but this Bf 110 did have the same smooth black finish and RLM 77 code colouring. The extended rear end of the fuselage, housing a dinghy, can be seen more easily in this view. The NJG 1 emblem can be seen in place beneath the cockpit.

Perhaps at first viewing this Bü 131, SA+ZV appears an unexpected addition to NJG 1's line-up of aircraft. The Nachtjagdgruppen's diving eagle emblem can be seen on the cowling. While similar to the emblem of NJG Schule Ingelstadt, there are clear differences, the latter being broader, the eagle's wings fuller and the letter 'S' appearing in the right upper corner. SA+ZV appears then to be a hack machine, used for communications flights by NJG 1 (a not uncommon practice amongst the Stab of many units). Overall colouring was 02 and, as specified in the RLM directive, the Balkenkreuz on the wings were aligned at right angles to the fuselage centre line; however they had incorrectly proportioned white angles.

A Fi 156 was more what one would expect in the way of a unit communications-cum-hack machine. CK+KF was in standard 70/71/65 camouflage and, as a second-line military aircraft, had retained its Stammkennzeichen (marked also in full beneath the wings). The NJG 1 emblem can be seen on the engine cowling.

When V.(Z)/LG 1 was redesignated I./NJG 3 on 1 October 1940 and transferred to night fighting duties, it initially retained its old code of 'L1' as seen here, marked in RLM 77 on the all-black finish of its Bf 110s. This was eventually amended to 'D5' while the old code was retained by the original parent unit. These Bf 110s belonged to 2./NJG 3, but no distinguishing colour was used in the code group, or Gruppe identification colour on the spinner tips.

In August 1941 I./NJG 3 was transferred, via Greece, to North Africa, with the intention of countering the RAF Wellington night bomber raids. However, it found itself employed mostly on daylight reconnaissance and strafing sorties, the lack of ground radar support facilities making night interceptions difficult. It returned to Europe at the end of October, being replaced by the Ju 88s of 2./NJG 2. This photograph was taken at an unknown location (note the He 111 in the revetment behind) in North Africa during gun alignment tests. Coded L1+DL, the 'D' was thinly outlined in the Staffel colour of yellow and the spinner tips in the I. Gruppe colour of white. The metre-wide white tactical band marking around only the upper half of the fuselage was characteristic of this unit, although it is a mystery why. Overall finish was the smooth nitro-cellulose black.

A brand new Do 217 J-2, GE+EA, fitted with FuG 202 Lichtenstein BC radar, in full night fighter black overall finish. The Stammkennzeichen was in 77, an interesting variation (they were often marked in white), and probably done as a concession to the colour used for unit markings on night fighters at that time. This code colour may have been used so that the aircraft could, if required before being given its service markings, operate at night without compromising the camouflage. The last two digits of the aircraft's Werknummer were marked in white on both sides of the fuselage, each in the centre of the fuselage Balkenkreuz as well as on the nose section. The type made its operational debut in May 1943.

Martin Drewes (centre) and other crew members stand in front of one of the III./NJG 3 Bf 110s, which still carried the 'Shark's Mouth' emblem of their former unit, II./ZG 76. (III Gruppe had been formed from II./ZG 76 and 4./NJG 1 on 1 November 1941). Compare the pale zig-zag lines of thinned out white used in a rough mirror image of the teeth markings. (The latter were painted with a dark colour on some Bf 110s of the unit, as their stark white colouring compromised their night camouflage.) The darker colouring of the random lines underneath that colouring matches what can be seen of the undersurface colouring. The large emblem eventually was painted over with a wash of a darker colour to reduce the white areas, which otherwise compromised the night camouflage to some degree. The aircraft in the background, D5+NT, had its overall general camouflage broken up with lines of colour slightly darker than the RLM 77-colouring of the code letters. (M. Drewes via S. Santos)

While the rest of NJG 2 were equipped with Bf 110 C-2, C-4 and D-1 (minus the infamous Dackelbauch long-range tank fitting), I./NJG 1 flew Ju 88 C-2s converted from standard Ju 88 A-5 airframes. This probably accounts for the retention of the day bomber camouflage colouring, while the rest of NJG 1's night fighters were finished in overall black. This 1. Staffel machine, R4+GH, fitted with FuG Lichtenstein BC, had had its aircraft letter outlined thinly in white. The black code letters prove that the camouflage was not black. The overall airframe colouring matches the propeller blades and hub – namely RLM 70. The same colour was used on the long flame damper fitted over the exhausts, while the open wheel-bay door, gondola and wing and horizontal tailplane leading edges all show the same uniform dark toned colour.

It is possible that the overall colour was a uniform RLM 70, but that begs the question why the smooth black nitro-cellulose finish, in use with the rest of the I./NJG 1's aircraft, was not used. There is evidence of a period of experimentation with colours for night fighting (see Volume Two, pages 303 and 306), and this aircraft's camouflage may reflect that. The colour is closer in tone to RLM 70 than that of RLM 77, so it is possible that an overall finish of RLM 74 might have been used – but this is pure speculation based on the range of camouflage colours available at that time. The Balkenkreuz on the fuselage has been crudely oversprayed with temporary black. The pale oblong patch of colour on the fin is possibly an 02-primed repair.

This Ju 88 C-6 was finished in a very thin overall application of temporary white, leaving the standard yellow tactical markings of a band around the fuselage and colouring beneath the wing tips visible. The stubs of the four antennae mast-fittings can just be seen, radar being of little or no use on the Eastern Front. Few Russian aircraft operated at night and areas were so vast that the range of the airborne radar equipment was useless without ground station vectoring facilities, something also lacking. The addition of flame dampers and the line of very dark paint along the leading edge of the horizontal tail surfaces, point to the aircraft's nocturnal role. (S. Santos)

Two views of a brand new Do 217 N-2, GG+YG, awaiting delivery to an operational night fighter unit. This version entered service in mid-1943, which provides a good time marker for the camouflage details. The overall dark finish of the uppersurfaces was RLM 75, with 76-colouring for the lower surfaces. The 75-colouring has been extended down as a square to form the background for the Balkenkreuz on the fuselage. The overhead view from the rear shows that the wing uppersurface Balkenkreuz has been correctly applied, just the white angles being marked. The two thin lines at each wing-root were in red (they match the red of the gust lock fitted to the port aileron) as a warning not to tread on the inboard end of each wing. Note that the small panel fitted between the engine cowl and wing leading edge was in a paler colour, probably 02, as was the rear end of the port engine nacelle and the small hatch cover on the uppersurface of the fuselage, just aft of the Balkenkreuz. The air scoops on the engine cowlings were also still in their 76 colouring, small 'defects' which would be corrected by painting before delivery. FuG 202 Lichtenstein BC aerials were fitted to the nose cone.

This high-angle view of the cockpit of a Bf 110 provides some interesting detail rarely otherwise seen. The official camouflage directives called for RLM 66 to be used in cockpits and any area visible through glazed panels, as visually they were considered to form part of the external camouflage. Originally, all internal surfaces were required to be painted in 02, although many parts, such as cockpit flooring, often remained in bare metal for practical purposes. RLM 66 was in use for instrument panels at that time and it was decided to extend the use for all internal surfaces visible from outside, the dark tonal value being adequate to match with the then current camouflage colours. This machine appears to have been caught in the transition, because while all the internal framing visible was still in 02, RLM 66 had been used on the horizontal surfaces of the cockpit liner – but only as far aft as just behind the pilot. The cockpit walls in the rear area were still in 02, as was the remainder of the horizontal cockpit liner surface. Whenever the RLM introduced retrospective changes to markings and internal colours, inevitably there arose anomalies; which perhaps reflects the overly bureaucratic form of language used for official documentation during the 1930s and 1940s. The black curtains, suspended between the front cockpit and the radio operator's position, provide the clue, indicating that this was a night fighter variant. This accounts for the abrupt demarcation line between the 66-colouring and that of the rear part of the aircraft. Externally the camouflage appears to be a 74/75 scheme with dense mottling to the fuselage sides.

The He 219 V4 first flew in March 1943 and the V5 was completed shortly afterwards. No further definitive information has been found to date on the production of the V6 seen here, which like the V5 in an all-black finish. The V4 and V6 were both used for tests with the Heinkel-designed compressed air ejector seats for both crew members. This required alterations to the cockpit structure with the upper canopy structure being eliminated for the trials. An air-to-air photograph of DV+DI in use for ejection seat trials shows it in conventional night fighter camouflage of 76-overall colouring with 74/75 mottling on the plan view uppersurfaces. However, there are photographs showing both the V5 and V6 finished in black overall, so it seems unlikely that the V4 would have been given the camouflage described for DV+DI or repainted. Test aircraft were not usually subjected to such esoteric and time consuming changes. This suggests that there was at least one additional aircraft used for ejection seat trials. Several of the first of the A-0 series, which were also allocated Versuchsnummern, were attached to I./NJG 1 at Venlo for operational testing and were marked with that unit's regular codes. Arriving in the first week of June, the first operational sortie was flown on the night of 11/12 June 1943 by the Gruppenkommandeur, Major Werner Streib.

Ju 88 night fighters used a wide range of camouflage finishes. This close-up view of a Ju 88 G-7b, ??+BR of an unidentified 7. Staffel illustrates a base colour of 76 over which a mottle of 75 has been applied, then enhanced with a darker colour in a pattern of wandering lines. The effect is more intense on the visible portion of the starboard wing uppersurface. This darker colour may have been 83 or 81. The muzzle flash suppressor fitted to the rearward firing gun can be seen, as well as the offset weapons pack underneath the fuselage. Only one of the Schräge Musik pair of 20 mm MG 151 cannon was fitted at the time of this photograph.

This is the starboard fin from Heinz-Wolfgang Schnaufer's Bf 110 G (now stored at the Australian War Memorial). His complete wartime tally of 121 kills is recorded on this and also the port fin (held by the Imperial War Museum, London). They commence with one marked 2 June 1942 and the last three are dated 7 March 1945. The fin, like much of his machine, had had the 76-colouring toned down with a thin spray of 02, producing a slightly yellowish tint, although the passage of 60 years has no doubt contributed to the colour change. Where the 02 coating has flaked off, the original 76-colouring can still be seen and is in good condition. (A.W.M.)

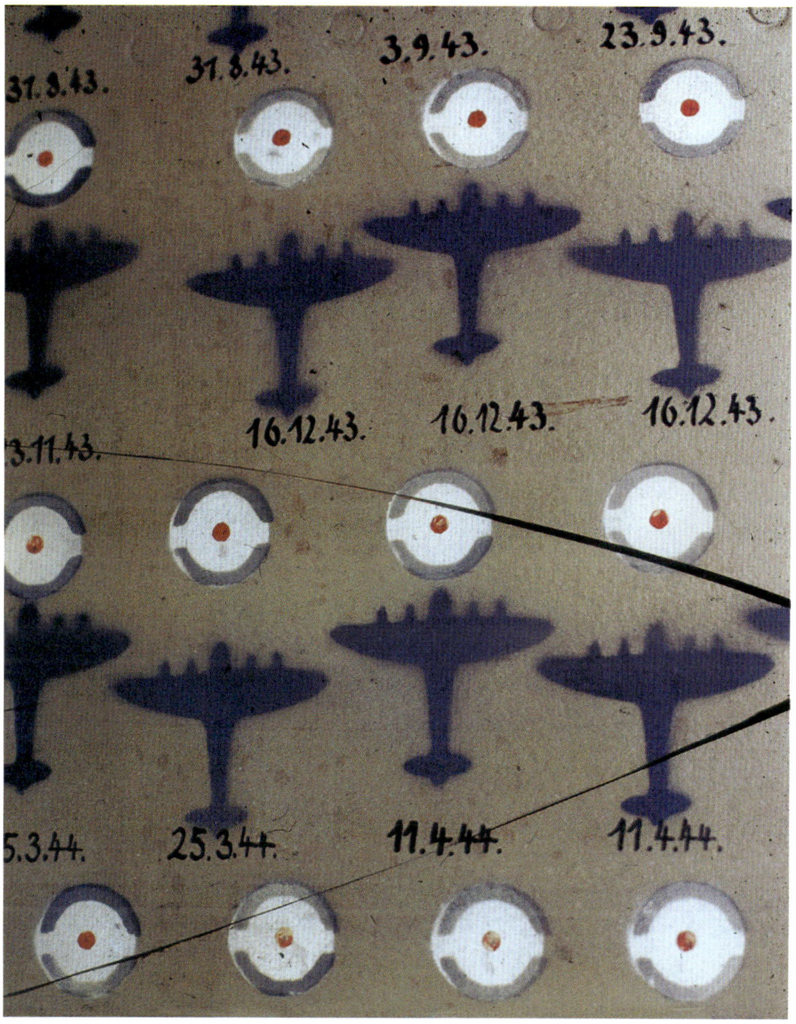

ABOVE: A close-up view of some of the markings that were applied using a stencil for the aircraft symbols, marked in 75, with the dates hand painted in black. The roundel markings were applied using two stencils, one producing a white disc, the second to produce the blue outline and red centre dot. (A.W.M.)

The last new night fighter type to see operational service was the Me 262 B-1a/U1, four of which formed part of seven Me 262s which equipped 10./NJG 11 (see Volume Two, pages 312-314). These four machines exhibited some distinct variation in camouflage despite all four having been converted and painted at the DLH workshop at Berlin-Staaken. Seen here, 'Red 10' W.Nr. 110635, was finished overall in 76 with mottling in 75 on its upper and side surfaces. Distinct differences in the style of application between port and starboard are obvious – the consequence of two painters, one each side. The wing uppersurfaces were a mottle of 75 over 76 but the horizontal tailplane uppersurfaces were in solid 75. The undersurfaces of the aircraft were in permanent black nitro-cellulose 7160.22. The red '10' was thinly outlined in the Staffel colour of white. Hakenkreuz and Balkenkreuz were black outline form on the fuselage and wing uppersurfaces. Drop tanks had been given a rough coating of black.

Ground-Attack Aircraft

Despite its 1935 origins, the Hs 123 was in the front line force of the Luftwaffe, the machines of II.(Schlacht)/LG 1 carrying out the Luftwaffe's first ground support sorties of the war. This Hs 123 A-1, from Sch.G. 1 was serving on the Eastern Front at the time of this photograph. Camouflage was 70/71/65 with its black Schlacht triangle marking outlined thinly in white, as was the black letter 'B'. The width of the tactical yellow band around the fuselage was oversize – but very effective. The time period was around Spring 1942. The Balkenkreuz on the fuselage was marked without the narrow black edging, and the Hakenkreuz was smaller than standard (note the sheen on it produced by the enamel paint used, compared to the matt surface of the camouflaged surfaces behind it). The white Infanterie-Sturmabzeichen close-support emblem can be seen directly beneath the inter-plane struts.

ZG 1 also used a large emblem for its aircraft, as shown here on this Bf 110 E fitted with an ETC 500 dual bomb rack. While the spinner tips were in yellow, the rest of each spinner appears to be green when compared with the colours of the Wespen emblem, indicating a III. Gruppe Stab machine. The location was the Eastern Front. This unit's practice of placing the yellow tactical band well forward of the Balkenkreuz on the fuselage allows just a small portion to be seen above the port engine. The camouflage is quite dense and there are several small darker patches on the side of the fuselage beneath the cockpit area.

Bf 110s of the Ergänzungs Schlacht Gruppe carried a huge unit emblem in the form of a black wolf's head, which must have done little to help the camouflage. This example, 4M+TB ('White T'), gives some impression of the sheer size of the marking. Spinners were marked in standard one-third white/two-thirds black proportions. The aircraft appears to have the 71/02 with 02 sprayed down the sides of the fuselage. Two SC 250 bombs were fitted to the ETC 250 bomb rack. (S. Santos)

The role of the Ju 87 had changed by 1942. Units on the Russian Front equipped with the Ju 87 D were being used in the ground support role and facing increasing opposition, resulting in rising losses. As a result, the Fw 190 gradually took over this role beginning in early 1943. However, this changeover was drawn out, so the Ju 87 lingered on in its daylight role for a considerable time before also being used in night ground-attack duties. Although this Ju 87 D-2 was fitted with a glider-towing attachment, it had been issued to a front line unit, indicative of the loss/replacement needs during 1942. W.Nr. 2705, T6+SB (the last letters of Stammkennzeichen were XU), belonged to the Stab of I./St.G 2; the use of a white numeral on the wheel spat is possibly indicative of its relative Stab position, white also being the I. Gruppe colour. The spinner tip was green. The usual Eastern Front yellow tactical marking was used under each wing tip and as a band around the rear of the fuselage.

This II./ZG 76 E-2 fighter-bomber, M8+LM, retained the white tactical marking introduced for the month of September 1940, confined to just the removable upper panel of the weapons bay on the Bf 110. The ETC 500 bomb rack can be seen beneath the fuselage, as well as the wing-mounted pair of ETC 50 bomb racks. This variant had a strengthened undercarriage and DB 601 N engines, indicated by the use of a white letter 'N' on each engine cowling. Camouflage was 71/02/65 with 02 sprayed as a soft mist and then 02 and 71 mottling used to tone down the colouring of the fuselage sides. The spinner tips were white.

The terrible conditions on the Russian Front were never conducive to any aircraft fitted with spats so all types rapidly dispensed with them, as seen here on this Ju 87 D-1, S2+NM of 4./St.G. 77. The choice of the Staffel colour (white) for both the aircraft letter and the spinner tip was unusual, the latter usually reflecting the Gruppe colour which, in this instance, should have been red. The wear on the 70/71 camouflage was indicative of the very harsh conditions (freezing winters and scorching summer heat) faced by the Luftwaffe, its aircraft relegated to an outdoor environment for their operational life. The black ‘S’ of the code group was worn, and the ‘M’ has a peculiar, broad form that does not match the style of the remainder of the code group, perhaps indicating an aircraft transferred between Gruppen. If so, the letter ‘N’ must have been retained, as there is no evidence of repainting. That may also account for the white spinner tip colouring. The usual yellow tactical markings were present beneath wing tips and around the fuselage.

This close-up view of the Ju 87 D or G under maintenance provides some details pertinent to that type. The rear end of the flame damper exhaust pipe, fitted for night ground-attack sorties, can be seen. Just above is the circular hatch cover for the oil filler point, identified with ‘35 Ltr Schmierstoff’ stencilled in white. To the right is the brown triangular ‘ASM Rotring Intavia 100’ emblem, edged in white, which identifies the oil type. Marked on the externally fitted armour plate is the white-edged 04-yellow triangle marked with E 87 in black to call attention to the fuel point, the white arrow pointing to the filler point cap set into the wing surface. Camouflage was 70/71/65.

The markings system for ground-attack aircraft was revised in April 1943 (see Volume One, page 257) but in October the entire force was revamped. Between those dates, the single letter identification style used by the Fw 190 force was replaced by a numerical system identical in most respects to that of the day fighter force. This St./SG 10 Fw 190 F-8, 'Yellow K', marked with the white-edged black chevron of I. Gruppe Stab, At first glance it looks as though the whole tail unit has been painted white. However on a closer look there is every indication that the aircraft has been repaired at some point at a major repair facility where the whole tail assembly was replaced by a sub-manufactured unit (see identically painted tail units in top photograph on page 26). The camouflage used by some SG 10 Fw 190s was very dark, as seen here. The black outline-form Balkenkreuz on the fuselage is hard to distinguish. Adoption of the non-standard overall dark colouring no doubt related to the ground-attack role, where such colouring would have been a distinct advantage low down against hills and trees. Other Fw 190s, but not all, from St/SG 10 were also painted in the same dark colouring. Retention of the black outline form of Balkenkreuz was contrary to RLM regulations, but made sense in terms of the camouflage purpose in this instance. The fact that the dark colour was on the aircraft before the white empennage colouring had been added; the date of use of the white marking system; the presence of a letter-form of identification, and only dark greens 70, 71 or 80 being available, narrow the possibilities of colours that might have been used. Luftwaffe Maintenance Units (and this extent of repainting would have required such facilities) were provided with the range of relevant camouflage colours. Careful examination of the photograph reveals what seems to be a single colour, so probably it was 80. The presumption that this aircraft's Stab status was the link with the white empennage, however, is challenged by a further example. An Fw 190 F-8/R-1, 'Black 2' of an unidentified III. Gruppe operating in Czechoslovakia, also had a white empennage – indicating perhaps that the white marking was employed for a specific sector. (JaPo).

RIGHT: One of the Latvian crewed, obsolescent Gotha Go 145s of Erg. Nsgr. Ostland used for night harassing attacks against Russian forces. These aircraft were culled from former training units and retained their basic camouflage, in this instance 02-colouring overall. The unit emblems of their former owners were painted over, but the distinctive shape on this machine, as well as its positioning, indicates FFS A/B 72 (see Volume Two, page 319). What is interesting about this particular aircraft is the marking of the '6X' code as an outline form, using a stencil, but over-size, being approximately three times the specified standard height. (B. Rosch)

Grounded by fog, the Hs 129 Bs of 13.(Pz)/SG 9 had to be destroyed when Russian forces broke through to Wagrowiec (Tonndorf) airfield on 22 January 1945. Almost all of the Staffel aircraft were concentrated on the airfield at the time and had to be blown up so that they would not fall into Russian hands. W.Nr. 141537, seen here being examined by two Russian soldiers, had had its 70/71/65 camouflage streaked with solid lines of temporary white, obscuring the Balkenkreuz on the wing uppersurfaces. The yellow tactical markings were a band around the rear of the fuselage and the underside of the wing tips, painted inboard to the line of the second aileron hinge. The Stammkennzeichen letters were marked on the undersurface of the wings, in black, but only 'TQ' under the port wing is known.

Although quite well known, these photographs are still the best examples of Fw 190 F-8/R1 variants of SG 2 in Hungary, taken during the winter of 1944/45. The RLM 04 yellow 'V' tactical marking under the wing was introduced on 25 September 1944 for all Luftflotte IV aircraft operating on the, then, Rumanian Front (see Volume Two pages 294-295). It should have been accompanied by a 40 cm-wide band around the front of the fuselage, forward of the wings but, as the side view reveals, this took the form of a band aft of the Balkenkreuz (possibly the existing Eastern Front yellow band had simply been reduced in width). Also, just visible from this angle is the all-white tailplane, the dividing line following the fuselage-empennage vertical join line. Use of this marking for Schlacht aircraft is enigmatic, originally having been restricted to day fighters operating on the Western Front, from approximately July 1943 to March 1944 (see Volume Two, pages 285-287). Judging by the fact that the underside surfaces were fully camouflaged rather than left in bare metal, indicating a production date before the changeover to 81/82/76, camouflage appears to have been 74/75/76. The outline-form black Balkenkreuz beneath the wings and on the fuselage were correct for the period. The spiral spinner had been introduced for aircraft of IV. Luftflotte from 20 September 1944, several months after its official introduction for aircraft on the Western Front. The lower portion of the undercarriage fairings had been removed to alleviate problems from snow and ice. The front of the engine cowling on '< 2' appears to be a very dark colour, and the aircraft is carrying an AB 500 container, housing anti-personnel bombs, under the fuselage, and a camera has been fitted under the inner port bomb carrier instead of four additional 50 kg bombs under the wings, as in the aircraft behind. The exact identification of late war Fw 190 variants is complicated as many changes were made and some designations were replaced by new ones as time went by. The sequence of these events is long and complicated and may well be discussed in detail in a future publication. (H. Obert)

Photographed just weeks before the end of the fighting, these two Fw 190 A-8s from an unidentified 6. Staffel of a Schlachtgeschwader serving on the ever-shrinking Eastern Front, carry the final style of tactical markings promulgated in March. However, the width of the yellow band around the engine of each aircraft was at least double the specified 50 cm. Placement was also different on each machine, that of the aircraft to the left being further forward. The yellow-painted rudder of each was standard, there being little that could be misinterpreted for that part of the marking instruction. The camouflage appears to be fairly pale, too pale for 81 and 82, a factor highlighted by the large area of 81 or 83 that had been used on the machine on the left, so it is probably 74/75/76. At first glance it appears as if an old marking has been painted out on the aircraft to the left, but the shape and size of the marking on the rear portion of the fuselage is too extensive, and the same colour appears again in a wavy edge style, on the port undercarriage oleo covers. Note that the darker colour has been painted by hand, there being no distinctive soft-edge bleed characteristic of a sprayed application. Addition of camouflage to the underside of the aircraft in this area would accord with the ground-attack duties. This may identify the unit as St./SG 10, which operated from Czechoslovakia in the closing stages of the war, and had other aircraft with similar additions to the camouflage. The second aircraft seen here, however, exhibits no similar additional camouflage, either to its sides or undersurfaces, although there are traces of what appear to be straight strokes of some sort on the portside undercarriage oleo cover. (J. Crow)

The aircraft graveyard at Ceské Budejovice also contained the remains of several ground-attack aircraft amongst the mass of machines. 'White 4', in the middle, shows the same very pale camouflage with extensive areas of dark paint added to break up the starkness of the general camouflage, extending in this instance to the entire uppersurfaces. The additional colour is more like 83 than 81, judging it against the black of the Balkenkreuz. The very pale finish on these machines seems to be a result of the much reduced camouflage of just 76-overall with segments of 75, a camouflage combination tested first in 1943, and introduced in limited form in early 1944 (see Volume One, page 111). Here, the 76-coloured areas have been overpainted with the darker colour, thus reversing the tonal pattern. Compare the wing pattern on 'White 4' with that of the machine immediately in front of it. The yellow tactical band had been confined to just the width of the engine cowling ring. The rudder is also yellow and the spinner has a regulation white spiral.

'Black 2', W.Nr. 63801121, an Fw 190 G-8 from St./SG 10 photographed at Ceské Budejovice, shows a similarly styled patch of very dark colour that has been painted over generally pale camouflage, which appears to be just 75 over a base colour of 76. Note the soft mottle patches along the rear edge of the starboard wing. Again, the patch was confined to the side area, not added to the uppersurface camouflage and its shape did not accord with the painting out of a previous marking. The aircraft carried the same style of oversize yellow band on the cowling as well as a yellow rudder. (JaPo)

The aftermath. This group of three Fw 190s, formerly deployed on the Eastern Front (only a relative distinction in the closing months of the war), display the variations that inevitably occurred in the application of tactical markings. The two closest aircraft have the regulation 50 cm yellow band applied set back from the extreme front of the engine cowling; the third machine had its marking around the front edge of the cowling, using the naturally occurring width of the curved section of the cowling to define the width. The actual final regulation of 7 March 1945 stated '50 cm wide yellow band around forward section of engine.' It was all a matter of interpretation, and unfolding events gave little cause to examine finer interpretations. The middle and last aircraft both have their rudder painted in yellow (and presumably so does the nearer aircraft), which accorded with the final directive. Note also the variation in interpretation of how the white spiral was applied to the black colouring of the spinner. The first and third have almost identical white spirals, while the middle machine has a much thinner and angled spiral. The Balkenkreuz and Hakenkreuz on 'Black 5' (W.Nr. ???181), however, were absolutely correct for the period. Camouflage is difficult to be certain about. The distant machine had had its engine cowling replaced (which may explain the variation in positioning for the yellow band), the 76-coloured side surface and lower panels hanging open do not match the darker tone of the side area of the rest of the fuselage. The camouflage of the uppersurface of the cowling also does not match that of the rest of the fuselage. Under extreme magnification there are two colours present in the standard pattern, possibly 74 and 75. The darker uppersurface colour of the rest of the forward section of the fuselage appears to be 82 or 83. The side of the fuselage, where visible, shows a darker colour taken right down to the lower line of the fuselage, possibly in 82. The middle aircraft has what appears to be 76, tempered with an over-spray of 02 and mottled lightly with 82, with some 81 on the fin. The upper, segmented pattern appears to be in 81 and 82. (J. Crow)

Another war trophy, Fw 190 G-8, 'Yellow 2' W.Nr. 586875 from an unidentified II. Gruppe that had surrendered at Neubiberg. Note the contrast between the background colour of the fin and that of the fuselage side surfaces. The latter had had the very pale greyish-white 76-coloured surfaces toned down with a very thin spray of 02. The discontinuity between the colouring on the fin leading edge and the joint where the empennage met the fuselage emphasises the problems increasingly encountered in the late war period of matching pre-painted sub-contracted components with those of the main camouflage. The optical enlargement effect produced by the outline form of the Hakenkreuz is interesting. The tactical markings of a yellow band around the engine (marked approximately twice regulation size, but more and more common to a large number of Fw 190s and Bf 109s in the last weeks of the war) and yellow rudder are apparent; the latter looks darker because of the sun angle to its flat surface. The relatively low contrast between the camouflage colours indicates an 81/82 finish. The spinner had the regulation white spiral. In the background 'Black 21', a Bf 109 G of II./JG 52, had much higher contrast camouflage colouring, with dense, large mottling in the same two colours to its side surfaces. The wide yellow band around the front of the engine cowling emphasises the contrast in tonal value. (J. Crow)

Bomber Aircraft

The very graceful Do 17 V8 was flown in the 1937 contest held in Zürich, its competition number '10' painted in black on a white panel each side of the fuselage. Overall colouring was RLM 63 and the full complement of military markings was retained. Dornier retained this machine for development work. (J. Radziwill)

The Do 17 K was built expressly for export (see Volume One, page 203) and the Royal Yugoslav Air Force purchased rights to manufacture, but only two were built before the German assault on the Balkans. This one (either No 3348 or No 3363) was photographed in Egypt during April 1941. Yugoslav documentation identified I.G. Farben as the supplier of the camouflage paints used for the type, which consisted of a base colouring of Laubgrün (sage green), Pompeianisch (dark brown used as patches), Zug Grün (train – locomotive? – green), and Gelb (yellow). The colour descriptions are difficult to assess from their written form, but sage was a grey-green (63 – used here as the undersurface colour), locomotive green – a darkish green (62), while the dark brown description would approximate the role of 61. The yellow was the odd one out. Overall though, from the scant information available, the remainder were close approximations of the Luftwaffe colours. If the information that I.G. Farben was the paint supplier – and that is logical given the high standard of German paint chemistry of that period – then the colour pigment standard source would have been taken from the RAL system. And the RAL system was the source of the RLM's parallel numbering system for 61, 62, 63 and 65 colours (see Volume One, page 36). The aircraft type description was marked in black across the base of the fin, with the air force eagle emblem near the top, and the RYAF tricolour of blue (upper most), white and red on the tip of the rudder. The apparent paler colour of the nose section was produced by the angle of the light source on the relatively sharp angle of taper of the fuselage, forward of the cockpit.

Three Do 17 Es, the nearest coded 27●17, of K/88 in flight over mountains in Spain, all wearing standard 61/62/63/65 camouflage with the white and black Nationalist cross tail markings that were so effective, even at a distance.

ABOVE: Do 17s delivered to Spain in the closing stages of the war were camouflaged in the revised colours of 70/71/65. This example had had its Legion Condor (Spanish) markings added to the wings and tailplane, but there is no indication of the black disc marking on the fuselage, or any codes. The markings process had not been completed when the photograph was taken. (J. Radziwill)

This line up includes two Do 17 Ms (the second possibly 27●32) in 70/71/65 with the most distant one, a Do 17 E, in the old 61/62/63/65 colour scheme. (J. Radziwill)

An early production Ju 87 B demonstrates the Überführungskennzeichen system, which used the first two letters of a four letter code to identify the manufacturer (or licence producer), plus two to record the aircraft sequence in a production batch (see Volume Two, page 360). The first two letters, marked beneath the starboard wing tip inboard of the Balkenkreuz, were 'JU' (for Junkers) in this instance. Note that the markings had been applied only to the underside of the wings, and not to the fuselage, as per the RLM regulation of 28 September 1937.

This view from below of 54+B12 of 2./KG 255 shows that the temporary red lacquer finish used over the national markings during the 1937 war games exercises was applied to all six positions (see Volume Two, page 260).

The He 111 P coded A1+CC of Stab II./KG 53, seen on the right, was photographed in France during the winter months of 1940/41. There are two oddities in this photograph; first, the individual letter 'C' appears to have been applied in white rather than in the Gruppe Stab colour of green; secondly, the Balkenkreuz marking on the fuselage had been incorrectly marked. The fault appears to have occurred through attempting to simply enlarge the pre-October 1939 style of marking without increasing the proportions. Interestingly, the under wing markings of the Balkenkreuz are of the correct form and proportions. Such anomalies were frequent during this initial period when front line aircraft had to have their nationality markings changed by ground staff (see volume Two, pages 436-439). The positioning of the Hakenkreuz marking shows that this was a post January 1939-produced aircraft.

The Dornier 17 Z was the last, and most important, of the Do 17 development line, first appearing in 1938. It formed the bulk of the Luftwaffe's Do 17 bomber force by December 1939. This example wore a crisp factory pattern 70/71/65 camouflage application, with a somewhat faded '702' (the last three digits of the Werknummer) painted below the cockpit in a pale colour, probably white.
The Stammkennzeichen, marked in black on the fuselage and beneath the wings, was ?C+AQ (possibly BC+AQ). The style of the upper and lower surface camouflage division, with the Hakenkreuz marked across the two and centred on the fin, indicates that this aircraft was produced at the change-over point when the Reichs und Nationalflagge marking was reduced to just the Hakenkreuz in January 1939.

An He 111 H-2, T5+BU of Wettererkundungstaffel (Ob.d.L.) – meteorological flight, so essential to bombing operations – displays its 70/71/65 finish and the intense staining from the exhausts. This area of black became significant in formulation of the under wing areas specified for permanent night camouflage introduced in 1941. The aircraft's code has been applied incorrectly; it should have been centred over the painted-out Stammkennzeichen letters, HP+AO.

The overall finish of this He 111 P, 5J+FL of 3./KG 4, was in very good condition, and it is hard to detect where the red banner and white disc had been painted out, leaving the Hakenkreuz in the pre-January 1939 position. The spinners were in the Staffel colouring of yellow, as was the aircraft letter, 'L'. (W. Berge)

He 111, 5J+GL, from the same Staffel (as at bottom of page 64), shows its 70/71 camouflage pattern. Note that the aircraft letter 'G' is marked in yellow on the outer portion of each wing. The position of the Hakenkreuz indicates that this was a post-January 1939 production aircraft. (W. Berge)

An He 111 of KG 26, fitted with a 500 kg bomb. The apparent pale coloured ring around the nose of the bomb was actually sunlight reflecting off the raised ring, fitted to slow the rate of penetration. Note the wooden board revetment surface in use. (F. Grabowski)

Temporary black lacquer 7120.22, for use by night bombing forces, had been introduced into the air battle over Britain during June/July 1940 (see Volume Two, pages 320-324). While the colour balance of this well known photograph of an He 111 H being prepared for flight is far from perfect, it does manage to convey the effectiveness of this temporary black finish. The wheeled vehicle was a mobile heater unit, the heating ducting to the starboard engine being in use at the time. The side panel that has been removed is on the ground in front of the wheel. The intensity of the exhaust staining can be seen. This would become a factor in the later development of day and night camouflage systems.

Bombing up a Ju 88 A-4 of 2./K.Gr.106 with an SD 1000, identifiable by its prominent red stripe on the 65-coloured body of the weapon. The code M2+AK is an enigma. The aircraft letter appears to be green and is definitely not in its Staffel colour of white. Camouflage was 70/71/65. The Balkenkreuz had been roughly daubed with temporary black to subdue the white sections.

This Ju 87 R, A5+CH, was photographed during the Balkan campaign and carried that campaign's extensive 04-yellow tactical markings on its 70/71/65 camouflage. Note that the I./St.G 3 emblem had been carefully retained on the yellow of the engine cowling. Use of the tactical colouring over the entire empennage suggests strongly that this is a Stab machine, such practice normally being confined to Stab machines for obvious reason. (Note that masking of the Hakenkreuz left the underlying green colour between the arms of the marking). Careful examination of the area aft of the yellow letter 'C', (the 3. Staffel colour, not green as would be expected) has failed to reveal the final letter; however, if the Stab and Gruppe identities are correct, then it should have been a 'B'. The aircraft letter is repeated below the wing tip and also would have appeared in similar position above the wing. (Trisch via J. Crow)

Most Ju 87 Ds reaching the desert conflict retained only their Mediterranean theatre of operations white band marking, not adopting the full North African suite of markings. This Ju 87 D-3, S7+EP of 6./St.G. 3, like most of the small number of this model to see service there, retained its European camouflage of 70/71/65. The speed of transfers to prop-up the ailing battle for the Axis meant that D- model aircraft were, at best, not expected to serve long term in North Africa, mimicking what had also happened with some of the Ju 88 units. As such, their limited concession to full tactical markings was deemed acceptable.

Instances of Ju 87 Ds arriving in North Africa in 78/79 tropical camouflage are relatively rare because by the time that the D model was being issued, the war in the desert was already moving into its final, destructive phase for the German forces. This example did not even have time to have its Stammkennzeichen, DH+RR, replaced with its unit identification. As noted previously, the white tactical band marking around the fuselage was typically the only concession to the campaign area on Ju 87 Ds. The rudder angle was sufficient to cause shadow from the sunlight, and the deep shadow thrown by the horizontal tailplane creates the illusion of a dark coloured empennage. (J. Crow)

When Operation Anton was launched on 11 November 1942, German forces swept into formerly neutral Vichy France in response to the threat posed by the Allied landings in Algeria. White tactical markings were used, both for bomber and fighter aircraft, in the form of a band around the rear portion of the fuselage and around each wing. Three Staffeln, 6., 7. and 8., are known to have been deployed from KG 2 for this operation and this Do 217 E, U5+LT (W.Nr. 4269?) of 8./KG 2, shows the effectiveness of the temporary tactical marking. This Staffel also marked its aircraft and Staffel letter, in miniature, on each fin, immediately above the Hakenkreuz. The sides and undersurfaces were in permanent black in this instance, these units having been employed on night shipping attacks.

This III./LG 1 Ju 88 A-14 operating from Tunis wears a finish of 78/79, with RLM 80 added as patches and lines to the fuselage and also probably the wings, including the underside of the leading edge of the wing root and portion of the bomb carrier. Rough, short lines of black have been sprayed over this. The lines of black were carried around the leading edge of the starboard wing to the join of the engine cowling. The spinners were one-third yellow/two-thirds RLM 70, identifying III. Gruppe, and a white tactical band marking was around the rear part of the fuselage, across the first two characters of the code. The 20 mm MG FF cannon fitted to the front of the gondola was used very effectively for anti-shipping attacks. The unit spent much of its wartime service operating in the Mediterranean.

This pair of Do 217s of II./KG 40 show the normal state of markings for the type. II./KG 40 had formed with the Do 217 E-1 variant for anti-shipping duties, under the control of Fliegerführer Atlantik, in March 1941. The very dark camouflage used by the KG 40 Do 217s appears to have been 72/73 rather than 70/71, combined with black lower surfaces, as seen, here for night shipping attacks. Code letters were painted in black, only the aircraft letter being thinly outlined in the Staffel colour (white in the case of these two 4. Staffel machines), and the Balkenkreuz markings were reduced by overpainting of the white angles, to just a black cross with very thin white angles. The full code was repeated in white miniature characters on the outer face of each fin. In the background was F8+AM, which mirrored the same changes and camouflage style, but had only the last two letters of the code group marked in white on the fin, the 'F8' portion being painted in what appears to be red, the II. Gruppe colour. Both aircraft have had the Hakenkreuz painted out. Note the aircraft in the foreground is missing its tail wheel doors.

Two Ju 88 A-4s of III./KG 51 above the Caucasus mountains. The aircraft letter 'D' of 9K+DS, in the foreground, is thinly edged in the Gruppe colour of yellow, with the Staffel colour of red being used for the spinners. In contrast, the 7. Staffel machine, 9K+FR, used a solid application of yellow for the aircraft letter, but followed the Gruppe practice of using the Staffel colour (white in this instance) for the spinners. The more usual practice was to use the Staffel colour for the aircraft letter (better for rapid recognition between aircraft) and the Gruppe colour for the spinners. The tactical marking of a yellow band around the fuselage and yellow for the lower surfaces of the engine cowlings and wing tips was correct for the period. Note also the yellow had been extended as a thin line around the extreme front edge of each engine cowling. At the start of the Balkan campaign, yellow was used more liberally around entire engine cowlings (see Volume Two, page 274) in accordance with the markings directive, but as the campaign intensified tactical markings became more restrained in the interests of camouflage which, in this instance, was standard 70/71/65. Both aircraft were marked with the III. Gruppe emblem below the cockpit outlined in the Staffel colour of yellow.

A close-up view of the quite stylish III./KG 51 badge. The background colour was blue, and the shield was edged narrowly in red. The edelweiss flower was usually rendered in full artistic manner, with shading to produce a three dimensional effect. When war broke out, KG 255 was redesignated KG 51 and its III. Gruppe emblem was adopted for the entire Geschwader.

The Do 17 Z had proved a very useful bomber in the early part of the war but by late 1941 its days were numbered, Ju 88s replacing it in most units. III./KG 3 was the last to use the type on the Eastern Front. During the winter of 1941/1942, 5K+HR, of 7. Staffel, had been given a coat of temporary white on the 70/71 camouflaged surfaces, leaving just the Balkenkreuz on wings and fuselage and the Hakenkreuz visible. The broad yellow band was still quite obvious though, something the Luftwaffe could afford during that first winter when opposition was relatively low.

F1+AS, a Ju 88 A-5, of 8./KG 76 looking fairly undamaged after what must have been a skilful forced-landing. The sprayed application of temporary white left none of the underlying 70/71 camouflage exposed, other than for the canopy framing, which would have been difficult to cover without problems with the glazing and the unit emblem. The white even included the rear half of each spinner, which on close examination appear to have been in the Staffel colour of yellow, leaving just the red II. Gruppe colouring. The underside of each engine's cowlings was in yellow, as also would have been the underside of each wing tip but there is no evidence that the yellow band marking of the fuselage was left exposed.

An He 111 H-6, ??+MB, from the I. Gruppe Stab, possibly of the same unit shown on page 72, although the unit code is obscured by the temporary white finish (the 'M' is marked in green and the 'B' in black). The permanent black painted area can be more clearly seen here. The aircraft behind is painted in the same manner. The array of 50 kg bombs, complete with screamers attached to the fins, provides some idea of a new type of operation to be carried out, probably against troop positions rather than major fixed targets. Colouring appears to be dark green. Both RAL and RLM colours were used for bombs, making it difficult to distinguish the actual colour at times. The ground starter trolley was usually painted RLM 66 overall.

For night operations, the 65-coloured sections under each wing could be overpainted in 7124.22 black, a permanent paint that could then be repainted with 7125.65 for daylight operations (see Volume Two, pages 323-324). This involved just the areas of 65-colouring outboard of the engines, and the rear portion of the fuselage, aft of the gondola. This use of special ultra-thin paints allowed repainting as the tactical situation required, up to ten times before removal of all the paint. This view shows the permanently painted areas of black and the 65-coloured areas beneath the wings and fuselage.

The precision with which the temporary white paint was added to some aircraft was sometimes in direct contrast with rougher applications that covered almost everything. This example was a combination of sprayed finish on the main portions of the airframe and hand painting around the Stammkennzeichen. The latter had been very carefully painted out to leave just a thin black edge. The Balkenkreuz also received this careful treatment, the brush strokes delivering a slightly duller finish than the sprayed areas.

In the opening phase of the invasion of Russia the Ju 87-equipped units had fared well. Facing little serious opposition, they had supported the bombing campaign with their precision attacks. However, by 1942 the type was under pressure from growing Russian opposition, and its role by then also had shifted to that of close-support for the ground forces, from which it began to suffer higher casualty rates. Its days as part of the main bombing force were limited and by the beginning of 1943 it was starting to disappear from the day role altogether in favour of the Fw 190 in the ground-attack role. This Ju 87 D-1 trop, S7+ZN of II./St.G.3, was photographed over the Balkans, carrying a supply container beneath each wing; fitted with a parachute in the rear section, this type could be dropped at up to 400 km/h to deliver a variety of supplies to ground troops. The aircraft camouflage was standard 70/71/65, with tactical markings of yellow beneath the wing tips and as a band around the rear of the fuselage. The aircraft letter 'Z' has been all but obscured by the yellow band marking, and the last letter 'N' is almost lost against the well-weathered camouflage. Note also the use of yellow along the leading edge of the fin for reasons not entirely clear, unless it was intended as a quick recognition marking in head-on engagements. The bomb container was probably painted in a RAL colour, such items often falling outside the direct control of the RLM colour system.

The changes to codes and their forms of marking extended to a practice that grew steadily amongst some of the Kampfgeschwadern units late in the war, namely of marking the last two characters of the code on each side of the fin in miniature. This example, from the fin of Ju 188 A, 3E+PH of 1./KG 6, shows the aircraft and Staffel identifying letters 'PH' marked in white, which appeared in the same relative position, each side of the black-painted fin (see Volume Two, page 330). (H. Taylor via F. Grabowski)

He 111 H-16, W.Nr. 8433, carrying an anti-German Hungarian delegation, landed at San Severo, in Italy, on 9 December 1944. The aircraft had only recently been sold to the Hungarian government. Its Stammkennzeichen, BT+KV, had been painted out and replaced with a Luftwaffe code of 2B+DC marked in black; the aircraft letter 'D' repeated, in black, beneath each wing tip. Hungarian military codes were of a distinctive style, so the code carried (with the first two characters, marked in the reduced form prescribed by the RLM), was that of Nachtschwarm Luftflotte 1, the original operating unit before the He 111's sale to Hungary. On the rudder (lost against the 70-coloured area of the 70/71 camouflage), was marked a red-coloured '4', approximately half the height of the Hakenkreuz. This probably related to its former Luftwaffe unit rather than Hungarian use. The Werknummer was painted above the Hakenkreuz in small white digits. The 'white marking', near the tip, was 02-colouring exposed where the 71-colouring had been damaged. Both the Hakenkreuz and Balkenkreuz were of correct form (outline only), and spiral markings had been added to spinners, per the instruction promulgated on 9 September 1944 for aircraft operated by Luftflotte 1 and 4 (see Volume Two, page 294). The camouflage pattern was standard Anstrich-Muster A (see Volume One, page 72). Interestingly, the rear view shows the slightly darker edge to the long 'splinter' of 70-colouring that touches the Balkenkreuz. The standard practice of first marking the edges of a camouflage area with the appropriate colour, then filling in with the colour, sometimes produced this effect under certain lighting conditions. (J. Crow)

Seen at Vaerloese before the surrender in Denmark, an He 177 A-5, ??+GN of an unidentified unit, photographed at great risk by a Danish farmer. (The aircraft had its port wing propped up with a wooden structure, blocking details of the unit code.) The front view shows the very clean, crisp, pale finish with the upper-surface colouring confined to the plan view of the aircraft. The second photograph shows that a soft mottle of one or both of the uppersurface colours had been used to tone down the pale-coloured side surfaces. The colours appear to be 65 with a very soft-sprayed finish of 70 and 71 on the top surfaces. The last two digits, '39', of the Werknummer were painted in black on the nose section.

The Ju 188 was another of the Luftwaffe's bomber force to utilise snaking (Arabesken) lines of pale colour to break up the dark bulk of its 70/71 camouflage. This captured example reveals a fairly sparse application of what appears to be 65-colouring used for the pale lines. Note what appears to be the Werknummer of the aircraft, possibly in white on the portside of the fuselage beneath the open cockpit window. (S. Coates)

ABOVE: Another Ju 188 A, with its Werknummer 0339 stencilled across the tip area of the fin, immediately above the white outline form of Hakenkreuz. The basic camouflage was 70/71/65, a fact established from the official correspondence of October 1944, exempting Junkers from having to use the new colours of 81/82/76 for this type (see Volume One, page 115). The snaking lines appear slightly darker in places, an effect produced by using very lightly thinned down colour, allowing the dark greens of the background colouring to influence the top colour. The other common practice for this type of camouflage addition was to use the spray gun set on a wide, flat spray and to then angle the spray gun. This made one edge quite strong in colour relative to the other, which simply faded out. No unit markings can be seen, even at high magnification, so this may have been a brand new aircraft, captured before it could be deployed to a unit. (J. Crow)

This He 111 H-20 had suffered collapse of the port undercarriage leg during landing, snapping the broad wooden propeller blades and pushing one strut of the undercarriage through the uppersurface of the wing. Note the long flame-damping exhausts, characteristic of the H-18 to H-22 series airframes. Camouflage was 70/71/65 but the code group, with the first two characters applied in small characters at that stage of the war, is not discernible, just the letter 'E' in white, repeated on the leading edge of the wing. The colour of the spinner tips indicates I. Gruppe and the white of the aircraft letter 1. Staffel, but that is all that can be safely surmised about its identity.

While of indifferent quality, these two photographs of a Ju 188 A-2 of 1./KG 2, found abandoned at Bron airfield, near Laon-Couvron late in 1944, illustrate the extent to which some bomber camouflage had been modified. During the so-called 'Little Blitz' (Operation Steinbock) launched against the UK early in 1944, the bomber force displayed a wide range of camouflage modifications, but this example is perhaps one of the more extreme (see Volume Two, page 329). The original 70/71/65 finish had large areas of either 65 or 76 sprayed onto both fuselage and wing uppersurfaces, probably done with permanent lacquer as the narrow white outline-form Balkenkreuz overlaps the adjacent pale area. Additionally, the reduced size unit code letters of U5 were also painted over the pale background, something that would not have been done with such precision had one of the contemporary temporary finishes been employed. All lower surfaces were black, probably permanent lacquer 7124.22, as there was little employment on the Western Front for use of the day/night role temporary lacquer system of 7120.22 and 7125.65 (see Volume Two, page 323). The rudder appears to have been a replacement as its single base colour of 70 (see Volume One, page 80) had been oversprayed with Arabesken of what looks like 02-colouring, something not seen elsewhere on the machine. The full code (U5+KH) was applied either in 77 or red, probably the former judging by the low contrast level seen on the port side view (where the 'K' appears to be more like an 'F', due to light catching a panel line that runs horizontally through the letter). Neither the Staffel nor Gruppe colour was red (the Staffel colour, white, was used on the tip of each spinner). The entire code was repeated in miniature on the starboard side of the fin, a common bomber unit practice in the late war period. Unfortunately, it is impossible to determine if it was repeated on the port side, although that was the normal practice. Note also that the Hakenkreuz had been overpainted on both sides of the fin. (J. Crow)

While much innovation had taken place with bomber aircraft camouflage in the last two years of the war, standard schemes could still be found as on this Ju 88 A of 13./LG 1, (but not the wreckage in the background). The colours were 70/71/65, but yellow tactical colouring was present on the engine cowlings. The rudder appears to be a dark colour, but this is the effect of the shadow caused by the sun angle. The 'white' wing tip was an unpainted replacement part, as was the oblong repair of bare metal on the engine cowling, all of which points to the rapidity with which this and other machines of LG 1 were abandoned in the face of the Allied advance. The spinner spiral markings are worth noting for their dissimilarity. The codes were in the late war form, with'L1' in one-fifth size, and the 'EX' normal size. (J. Crow)

This enlargement of the wreckage of L1+BX(?), seen in the background in the photograph above, makes an interesting contrast in camouflage colouring with that of L1+EX. While blurred, it still shows clearly the dark green of the rudder, overpainted with a winding line of 76 (it is too white in tone compared to the 65 of LI+EX), while the fin has been almost completely covered with a soft-spray of 81-colouring. The upper-surface of the horizontal tailplane, as well as that of the fuselage, shows a green and a grey colouring, probably 82 over the base colour of 75, although there are also small patches of sprayed 81-colouring discernible as a mottle on the tailplane surfaces. The remains of the engine, to the left, appears to have a white-coloured spinner. (J. Crow)

Another in the line of abandoned and partly destroyed Ju 88 As of LG 1, discovered by American forces. The camouflage colouring appears to be 81 and 82, judging by the brown colouring visible on the engine cowling and nose section as seen in the original, extremely well-balanced colour photograph. Again, the engine cowling had been painted in 04 yellow. Note the different style of the white spirals compared to the photograph opposite; each had to be individually applied by hand. (J. Crow)

Maritime Aircraft

Amongst the fledgling Heinkel company's early designs was the He HD 42b floatplane. D-2033, W.Nr. 373, in the foreground of this line-up of nine, was registered to the DVS (Deutsche Veerkehrsfliegerschule GmbH) Berlin, in April 1931. The overall finish is not silver (aluminium dope) as expected for seaplanes of this period. The sunlight was quite strong, as shown by the strong highlight on the curved surface of the propeller blade on D-2033, yet there is no corresponding reflection on the curve of the engine cowling. The floats also appear to be relatively dull, which is significant, because even where other colour was used on early seaplanes, floats usually were given a coat of aluminium paint, which was always a good sealing surface paint against corrosion, especially for a marine environment. The date precludes any of the relevant RLM colours, such as 63 and 02, but the general tone is similar and appeared on a number of types of the late 1920s/early 1930s period. It must be remembered that there was already clandestine planning in motion for a new air force for Germany. The black/white/red tricolour band across the starboard side of the rudder dates the photograph to before September 1935. The Heinkel company logo was marked in black, near the tip of the fin. In the background are two He 9 floatplanes in the same colour scheme.

He 51 B-2 floatplanes, the only single-seat fighter type equipped for maritime work, entered service with the Küstenjagdstaffel in early 1936. They served with Küstenjagdgruppe 136, which also operated a land-based Staffel of He 51s. The unit eventually was redesignated as I./JG 136. Colouring was 63 with treated natural metal for the floats, which had a clear protective coating and a black anti-fouling finish on lower surfaces up to the high water mark. The number '12' was in white thinly outlined in black.

The design of the He 114 dated to 1935 and, while it was obsolescent by the time that war broke out, it continued to give good service until the mid-war period, supplemented by the Ar 196. The V1 and V2 prototypes flew in 1936, but the next two each featured a change to BMW radial engines, the V4, W.Nr. 2262 D-IDWS, having a 960 hp BMW 132 K engine. Overall colour was RLM 63, standard for the time period, with black civil registration and civil national markings.

The Do 18 gave valuable service for many years and in, doing so, made the transition in colour from the pre-war 63 overall finish to full 72/73/65 camouflage, as seen here. Note the use of red, yellow and black tip colouring to each propeller blade, a carry-over from pre-war days when this form of warning marking (in the old Imperial colours) on rotating propellers was seen on some land planes (e.g. Do 17s), but more often on sea planes, probably because of the restricted areas of movement for ground handling staff. The white 'Z' was the aircraft letter, repeated from the code group.

Another valued, but oft overlooked, floatplane type was the He 115. Camouflage was 72/73/65 with the Stammkennzeichen, VF+UY, marked in white, the latter not an uncommon feature of Heinkel-produced aircraft wearing dark camouflage, the relevant RLM directive allowing such usage. (via Jessen)

He 115 C-1, 8L+IH of 1./Kü.Fl.Gr. 906, was photographed at Santahamina, Finland, on 8 October 1942. The unit was one of six Gruppen to operate the type in the multiple roles of mining, bombing, torpedo attacks, reconnaissance, and sundry coastal duties. While the design dated back to 1936 and production was halted in 1941, such was its versatility that production was recommenced in 1943. Old though the design may have been, it proved to be the most successful attack and reconnaissance type used for maritime duties. Camouflage was standard 72/73/65.

For night operations, the He 115 wore temporary black for its undersurfaces, as used by the land based bomber force. The temporary black finish of 2M+BL of 3./Kü.Fl.Gr. 106 was showing considerable wear along the leading edges of the wings and the nose area in particular. The vertical streak of black covered the visual warning for the propeller tip area marked on the fuselage. The white angles of the Balkenkreuz have been crudely daubed with spots of the black to discolour them, indicating that this photograph was taken prior to about April 1941 when RLM 77 was introduced for that purpose (see Volume Two, page 323). The yellow letter 'B' has also been painted out, somewhat roughly. The pale area of camouflage behind the unit code M2 is something of an enigma, as it is definitely much lighter than the 72/73 finish of the nose section, engine wing leading-edge areas and floats. The 3. Staffel badge can be clearly seen.

A Fi 167 A-0 coded TA+AA, one of twelve produced, photographed during trials at Erprobungsstelle Rechlin wearing an overall finish of 02, the basic undercoat applied to all metal aircraft. Erprobungsstelle 167 was formed with these pre-production aircraft and nine were transferred to the Netherlands in 1941 for sea camouflage development trials (see Volume One, pages 120-121).
(D. Vincent)

RIGHT: The BV 138 design originated with the Hamburger Flugzeugbau aircraft subsidiary of the Blohm und Voss ship–building company, the first prototype being designated Ha 138 V1. Aerodynamic problems eventually led to a complete redesign, the first of which, D-ADJE, appeared under the designation Bv 138 A-01. Overall colouring (other than the black along the side of the engine beneath the wing) appears to have been RLM 02, RLM 63, or a similarly toned RAL colour. Given that it was produced as a military prototype, circa 1939, it was probably 02. Note the military version of the Hakenkreuz, rather than the Reichs und Nationalflagge normally used in conjunction with civil registrations. (Franz Selinger)

The BV 138 A-1 version entered service in mid-1940, in the maritime reconnaissance role. The BV 138 A-04 then served as the prototype for the B-0 pre-production series, a strengthened version, the production version being the B-1. However, some of the BV 138 B-0 aircraft were converted for minesweeping duties (as seen here on ?B+GA), and issued to the Minensuchgruppe der Luftwaffe. Camouflage remained the standard marine colours of 72/73/65, the suspended degaussing ring being finished in the same colours. Note the series of coloured 'aiming' lines beneath and forward of the cockpit, used for aerial camera work. The only known code for the Minensuchgruppe is 3K, so this aircraft appears to have retained its Stammkennzeichen, military code and Balkenkreuz normally being applied to each boom.

The principal production variant was the BV 138 C-1, in which all the previous shortcomings had finally been resolved with 227 of the total production of 279 being this variant. The Jumo 205 D engines were retained, but the central engine was fitted with a four–bladed propeller of narrower width, which ameliorated vibration problems associated with using three identical three-bladed airscrews. The SAGr. 125 emblem can just be distinguished below the front turret. What appear to be two sea mines can be seen beneath the starboard wing root area. (D. Vincent)

An elegant photograph of a Do 24 showing that even under oblique and strong sunlight, the uppersurface colouring of 72/73 on the wings blended well with the ocean background. The type served multi-purpose roles from transport, escort, search and rescue to reconnaissance.

The military Fw 200, an adaptation of the original pre-war airliner designed for Luft Hansa's long passenger routes, served in multiple roles: reconnaissance, convoy shadowing, anti-shipping attacks and – occasionally – as a make-shift troop transport. The type had not been designed for a military role, and it suffered its share of structural problems – but remained one of the most successful anti-shipping aircraft used by the Luftwaffe. Ten pre-production aircraft, designated Fw 200 C-0, were delivered the month that war broke out – joining four Fw 200 Bs in transport roles with K.Gr.z.b.V. 105, while the other six, armed, were taken up by the newly created Fernaufklärungsstaffel and sent immediately into action to monitor British shipping in the North Sea. The unit was redesignated 1./KG 40 in April 1940 and its role changed to anti-shipping duties; it was then redesignated I./KG 40 in June 1940, re-equipping soon after with the Fw 200 C-1 variant. The I. Gruppe emblem of the world in a ring is shown in detail here, a photograph taken at Vaerness, in Norway, in 1940. The emblem is different to that carried by Fw 200 operating in the Mediterranean area in as much as the world map has the African continent largely obscured by the yellow ring, giving more prominence to the north and Atlantic area. In the background is F8+CH, with its aircraft letter 'C' narrowly outlined in the 1. Staffel colour of white. The Werknummer details were usually marked in white on the fin on the early model machines, as seen here, just below the Hakenkreuz. Two successful shipping attacks had been recorded as white ship emblems on the rudder. Note also the oversize black centre to the fuselage Balkenkreuz, a common characteristic of KG 200 Condors (under-wing markings, however, retained their standard proportions). This reduced the unwelcome effectiveness of the regulation white angles of the standard Balkenkreuz, which compromised the dark camouflage, the type often operating at relatively low altitude where it blended into the seascape. Camouflage colours, initially specified as 70/71/65 for the land-based duties variant, was changed to 72/73/65 once the type became dedicated to the maritime role.

This air-to-air photograph of F8+CS, a Fw 200 C-3 of 8./KG 40, (the aircraft letter 'C' narrowly outlined in red), demonstrates the effectiveness of the white angles to the Balkenkreuz, even after being reduced in proportion. The Gruppe emblem is easily distinguished. (via B. Rosch)

As noted earlier, the Fw 200 was a jack-of-all-trades, serving many roles in addition to its principal one. Here F8+HK of 2./KG 40 was being used to transport troops during the winter of 1942/43, possibly from its base at Trondheim. The aircraft letter 'H', was outlined in red. Detail of the revised proportions of the fuselage Balkenkreuz can be seen more clearly here. The black of the unit code letters 'F8' was almost lost against the 72/73 colouring, only the very slight sheen of the enamel-based paint used for the code, plus the slight change in light angle forward of the Balkenkreuz, making them visible.

RIGHT: There were other more specialised roles for the Do 24 as well. KG 200, even stripped of the mythology which has been attached to that unique unit in the last 50 years, was engaged in some unusual operations and used a variety of aircraft types. This damaged Do 24, A3+FB, bears the KG 200 code. Colouring remained standard 72/73/65 but the Hakenkreuz and also the Balkenkreuz on the fuselage and wings, had been toned down with a wash of darker colour, probably temporary washable black.

The other workhorse of anti-shipping actions and ASR duties was the Do 24 T. This example, KK+UP, wore the standard 72/73/65 finish plus a Mediterranean theatre tactical marking of a white band around the fuselage. (D. Vincent)

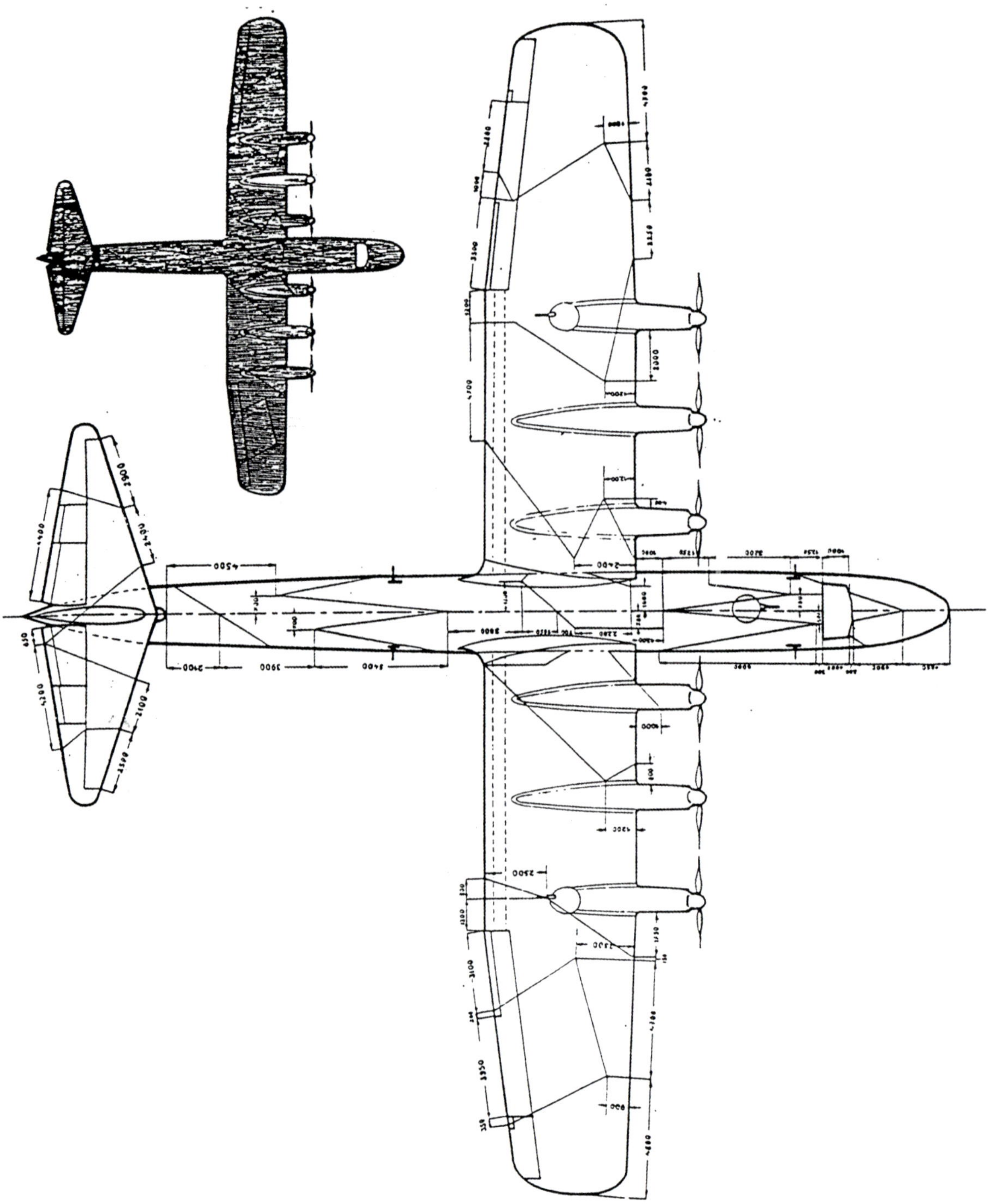

LEFT AND OPPOSITE PAGE: These are two diagrams of the official camouflage pattern for the BV 222 C series. The precise measurements for each segment were stipulated and a shaded diagram of each view was added to ensure that the camouflage was correctly applied. On an aircraft of such dimensions, it would have been difficult to spot any incorrect application without having such elaborate details laid down.

BV 222 C Allgemeine Angaben 0 21
Flgzg.-Hdbch.

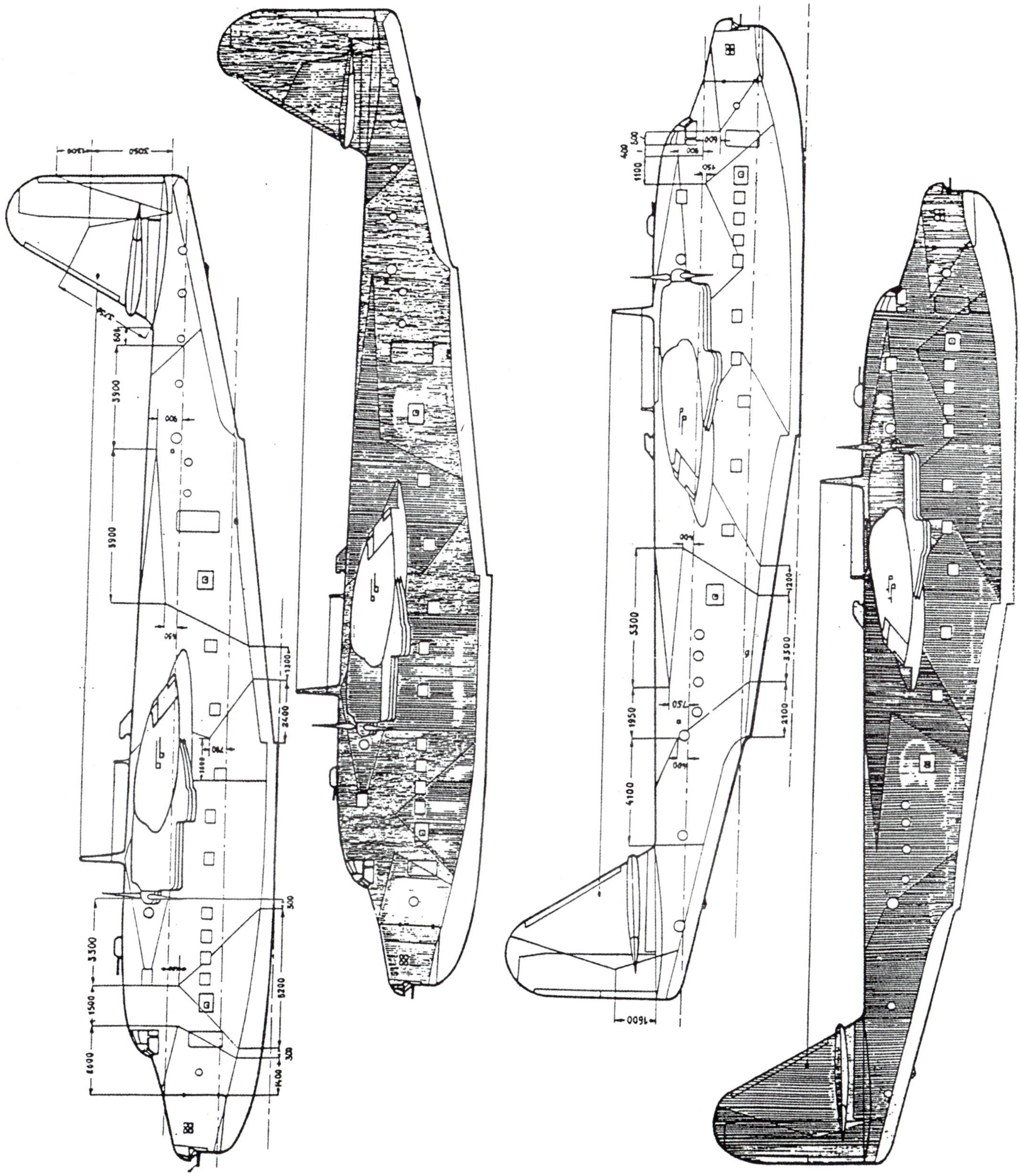

Abb. 5 : **Zweifarben-Sichtschutz**

The relatively strong sunlight and angle shows the pattern of the camouflage quite clearly on BV 222 V3, DM+SD, W.Nr. 439. Comparison with the camouflage diagrams reveals the care with which this giant aircraft type was painted. The wing tip undersurfaces were painted yellow, while the fuselage had a white band (this combination of tactical colours also being used on the V4 as noted further on). In May 1942, this particular aircraft was issued to 3. Staffel of Küstenfliegergruppe 406 (which had BV 138s), based at Biscarosse. (via M. Jessen)

BV 222 V8, X4+HH, the last of the A-series airframes, was also serving with Lufttransportstaffel See. The unit's Viking ship emblem can be seen aft of the cockpit. Again the photograph shows clearly the precise application of the 72/73/65 camouflage scheme. The code letters on the BV 222s of this unit are difficult to discern because they were always applied only in black. The aircraft was being prepared at the Blohm und Voss works at the time and the Theatre of Operations white band around the fuselage identified its intended operational area as the Mediterranean. However, close examination of this and other photographs shows that at least the V3 and V4 (both marked with a white band), V7 and V8 were painted yellow on the lower wing tip surfaces. In company with the V1 and V2 machines, the V8 and V6 were used for supply flights from Greek and Italian bases to North Africa, landing at Tobruk and Derna. The V8 was shot down over the Mediterranean on 10 December 1942 by Beaufighters, the V6 escaping with heavy damage.

LEFT: The BV 222 C-07, serving with Lufttransportstaffel See as X4+CH, again provides good evidence of the consistent and exacting replication of the camouflage pattern as set out in the accompanying diagrams. Given the size of this aircraft type, applying the entire pattern accurately would not have been an easy task. This was one of four Versuchs machines (V4 X4+DH, V5 X4+EH, V6 X4+FH) transferred to the Luftwaffe between 20 April and 26 October 1942.

In this overall view of the same aircraft, (opposite page bottom) the last letters of the code, CH, may be seen more clearly. The forward location of the Balkenkreuz on the fuselage was unique to the BV 222 series of aircraft. Interestingly, the subsequent BV 238 aircraft reverted to the more usual location for this marking.

Detail of the 65-coloured scribble pattern can be seen on this close-up view of the nose of this Ju 188 D-2 of KG 26, taken after the surrender in Norway. The scribble blends into the lower surface 65-colouring where the two meet, but takes on a slightly paler appearance where it is seen against just the dark greens of the background, an optical illusion caused by strongly contrasting colours. Uppersurfaces were 70 and 71. FuG Hohentweil 200 search radar is fitted. On the right is the fin and rudder of W.Nr. 0607, its 72/73 camouflage broken up with 'seams' of 65. Compare this extra application with that shown on W.Nr. 0608 in the photograph on page 90, top. The two aircraft in the background have contrasting camouflage. The example on the left has had so much scribble pattern applied to the nose section that it appears to have a pale finish with a few dark mottles. The aircraft on the right had its original dark standard camouflage taken down the side area of the nose section to the bomb-aimer's window and the Arabesken pattern was less intense.

A line-up of the same unit's aircraft (from previous page, bottom) illustrates the random application and style of the Arabesken pattern. Note the absence of this addition on the engine cowlings of the second and fourth machines. The unit code of '1H', marked in regulation small size, is almost impossible to discern, as intended for security purposes. The nearer aircraft is 1H+BD from Stab III. Gruppe, the next, W.Nr. 0608, is 1H+BR from 7. Staffel. On the ground further down the line is a long-range drop tank with which these aircraft were equipped. The fourth, fifth and sixth machines are coded 'FR', 'CR' and 'DR'.

More of the vast collection of Ju 88s and Ju 188s of KG 26 captured in Norway at the war's end. The nearest aircraft, a Ju 88, has streaks of a colour that appears to be 65, used in a relatively constrained form for this type of camouflage combination. Note the white spiral on each spinner has been confined to the outer section of the two-part spinner (it no doubt being easier to achieve this difficult marking with just the smaller element). The tip colour is yellow. Seen between its undercarriage legs are two other machines, a Ju 88 in conventional camouflage to the left, 1H+BR marked in black, and a Ju 188 D, 1H+BS, the latter with the additional camouflage. In the centre foreground is a Ju 188 D in standard camouflage finish, marked forward of the Balkenkreuz with 'B' in a pale colour that looks like green or yellow. The Balkenkreuz on the fuselage had been toned down with a wash, possibly of thinned down 02. Compare the broad spiral on each spinner with those on the Ju 88 in the left foreground. (S. Coates)

Reconnaissance Aircraft

The He 45 was originally designed as an unarmed light bomber but, by the time it reached production status, its usefulness in the anticipated role had lapsed. Instead, it was used for reconnaissance duties. This early example was finished overall in silver (aluminium) lacquer. (P. Cohausz)

A Do 16 Wal seaplane of the Seeaufklärungs unit based at List from late 1934. The registration D-ABAU was painted in regulation black but the strong sunlight makes the markings on the upper surface of the wing appear paler. The weights and loading table can be seen marked on the fuselage, just below the tailplane.

While the He 45 filled the long-range reconnaissance role, the He 46 served on short-range duties, being paramount in this role with the Luftwaffe in 1936, but rapidly disappearing once the Hs 126 began to enter service. In 1935, the He 46e was developed as an export version, D-IBKA being one of only 18 built with a cowled British Bristol-Siddeley Panther engine. Its finish had a definite gloss to it and the overall colour matches closely the pale grey of the Heinkel-produced airscrew blades. This may be the earlier, paler overall colouring referred to in Volume One (pages 25-26), or else standard RLM 02 that appears paler because of the glossier finish.

The internecine conflict in Spain provided opportunity to test the He 45c under battle conditions. The initial supremacy of the Legion Condor forces was rapidly diminished with the supply of aircraft from Russia, and soon made camouflage necessary for the slower German-manned aircraft. Supplies of standard camouflage colours of 61, 62, 63 and 65 were made available to the erection and servicing depots used by the Nationalist forces.
The He 45cs, along with several other types, were given a rudimentary camouflage, in this instance using all four colours, as some unique colour movie footage proves.
The 61/62/63 patterns used on upper and side surfaces were entirely random. The Nationalist Forces' combination of white markings to the wing tips and rudder with black cross, plus white crosses on the upper surfaces of the top wing and black on the lower surfaces of the bottom wing, were added. The playing card on the fin was probably a personal emblem, while the Devil emblem was possibly the Staffel marking. Each type used by the Legion Condor had its own code number, '15' being allocated to the He 45s, marked in black and separated from the aircraft's individual number by a black disc marking.

The He 45 lingered on in service, this example 10+A11 of Aufkl.Gr.(F)/121 being photographed in 1938. By that time the overall pale green-grey 63-standard finish was still retained on some older aircraft. As the war drew closer, the earlier finish was changed to camouflage of 70/71/65 when the type was still serving, albeit in small numbers, with Staffeln of five of the short-range reconnaissance units in September 1939.

The reconnaissance role specification produced a number of contending designs, including one originally intended for use on the aircraft carrier 'Graf Zeppelin' in the dual role of reconnaissance and torpedo attack. Prototypes appeared in two forms, fitted either with floats or with a fixed undercarriage, as seen here on the Ar 95 V4. The overall finish was RLM 63 (02 being used only for floats on water-borne maritime aircraft of the period). The walkways on either side of the fuselage were finished in black, the exceptional width necessitated by the steep slope of the inner wing section.

The predecessor to the very successful Hs 126 was the Hs 122, three prototypes appearing in the summer of 1935; the Hs 122 V1 (D-UBYN), V2 (D-UDIZ) and the V3 (D-UBAV). The Hs 122 V2, seen here, was fitted with a 660 hp Siemens SAM 22 B radial engine. Overall finish was probably RLM 63 given the type's military focus, registration markings being applied in black in all four locations. A standard weights and loading table was marked on the port side of the rear section of the fuselage. The cowling appears to be a dark colour, but not as dark as the civil registration markings, possibly a dark grey.
A small batch of pre-production Hs 122 A-0 machines was built, but the Henschel Company's hopes of gaining a contract for the type as a successor to the ageing He 45 failed, its performance being little better than that type.

RIGHT: The Henschel Company had more success with its next design, the Hs 126, which was based on the Hs 122. The new design made its maiden flight in the autumn of 1936, using the converted Hs 122 A-04 airframe. Development went well and ten Hs 126 A-0s were built and several evaluated by the Lehrgruppe of the Luftwaffe. The first production examples began to leave the Schönefeld production line in early 1938.
This Hs 126 A-1, HS+IS W.Nr. 3058, one of ten photographed while awaiting delivery at Schönefeld in the Spring of 1938, was camouflaged in the newly introduced colours of 70, 71 and 65 but retaining the Reichs und Nationalflagge across the fin and rudder.
The civil registration system had by then been abandoned for military aircraft deliveries, as of 23 September 1937 (see Volume Two, page 360), and a four-letter Überführungskennzeichen replaced it which utilised the relevant manufacturer's initials for the first two letters.

When it entered service with Luft Hansa in 1934, the He 70 design had made civil aviation history with its record-breaking flights. That same year the civil version was put into military service for high-speed communications services, and a small number were on strength when the Luftwaffe was officially announced in the following year. Two versions were used, the He 70 F-1 for fast long-range reconnaissance, and the He 70 E as a high-speed bomber variant. The three He 70 F-1s seen here were serving with Aufklärungsgruppe (F)/124 (later redesignated Aufklärungsgruppe 24). All three Staffeln of I. Gruppe are represented by this photograph, 40+K11 of 1. Staffel, 40+L12 of 2. Staffel, and 40+K13 of 3. Staffel. Overall colouring was RLM 63 combined with the black trim and lightning bolt marking of the civil version. The broad wing, which hampered the observer's work at some angles, and the overly sensitive flying characteristics were a problem and by 1938 the type was being phased out in favour of the Do 17 F.

The Do 18 design originated in 1934, the specification being issued to Dornier by Luft Hansa, which was seeking a replacement for the Dornier Wal for its maritime services. The first prototype flew on 15 March 1935 and the type underwent steady development before being taken onto the inventory of the Luftwaffe in 1938, the military variant Do 18 D becoming operational in September of that year. Issued to the Küstenfliegergruppen for long-range reconnaissance work, the type was spread amongst the various Gruppen, initially usually as a single Staffel. 60+D52 was serving with 2./KüFl.Gr. 506, based at Kiel under VI. Luftkrieskommando, when photographed. Maritime camouflage was still 63/65 at this time, and would remain so well after the land-based forces had abandoned the old 61/62/63/65 scheme for the revised colour scheme of 70/71/65. It would not be until the following year that maritime colouring was changed to 72/73/65 (see Volume Two, page 333). The Staffel's emblem was marked on each side of the engine cowling. The discolouration on the fin and rudder appears to be from engine exhaust staining. The absence of the red background and the central placement of the Hakenkreuz points to this photograph having been taken after 1 January 1939, at which date the old red banner and white disc portion of the Reichs und Nationalflagge had been ordered to be deleted on military aircraft (see Volume Two, page 436) in response to growing international tensions.

This pair of Do 18 Ds of 2./Kü.Fl.Gr. 306 (soon after redesignation of 2./Kü.Fl.Gr. 406), was photographed at Wilhelmshaven in October 1939, by which time the revised maritime camouflage of 72/73/65 had been adopted. The nearest machine, its aircraft identification letter 'D' marked in white outline form on the front of the engine cowling, had the 2. Staffel badge marked on the bow. Note the white pylon marked with a red band, of the beaching trolley and also the camera angle lines marked beneath the cockpit window as an aid for the hand-held camera used in reconnaissance work. The under-wing Balkenkreuz, while appearing out of proportion, was the correct size for the type. In the background, on the right, is an He 59 in full 72/73/65 camouflage, marked WL+AHD, the revised civil registration introduced on 1 January 1939, to signify second-line military status (see Volume Two, page 360). (F. Selinger)

The Do 26 also made the transition from its civil role as a non-stop trans-Atlantic mail carrier with Luft Hansa, to that of military reconnaissance at the war's outbreak, the Luftwaffe being short of long-range maritime reconnaissance aircraft. The Do 26 C (alias the V 4) P5+DF, which, with the V 1, V 2, V 3, V 5 and V 6 formed a Sonderstaffel attached to Kü.Fl.Gr. 406 for long-range reconnaissance duties in 1940. Camouflage was 72/73/65.

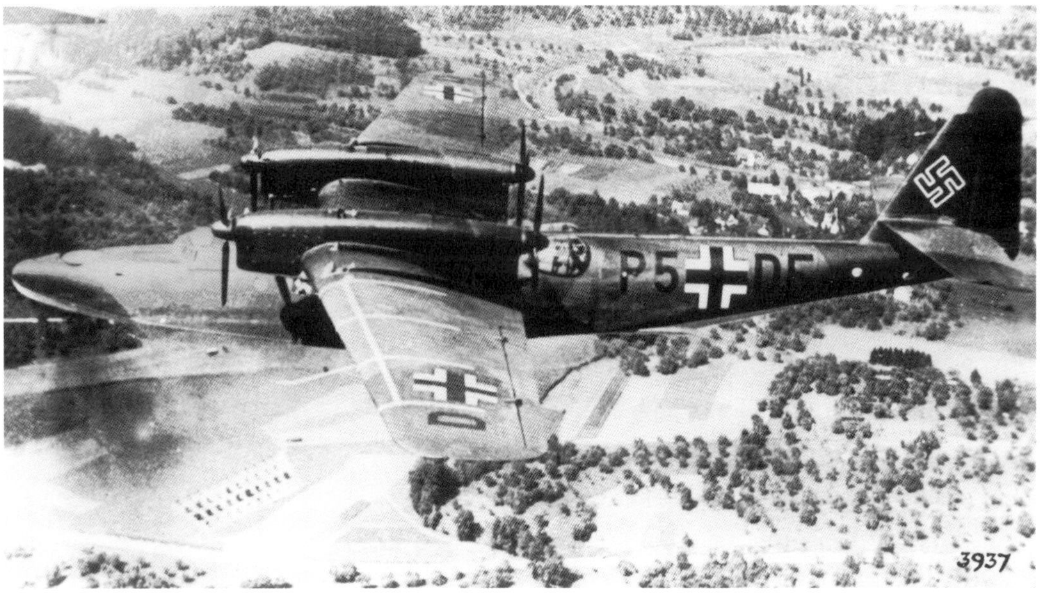

The BV 138 was another pre-war design that entered service at the war's start, albeit in small numbers, and initially in the reconnaissance role. Later variants had their role extended to include shipping attacks and minelaying. Its robust construction allowed for use in open seas and its maximum range of 5,000 km (without war load) was extraordinary. Camouflage on K6+GK of 2./Kü.Fl.Gr. 406 was the marine colours of 72 and 73 with 65 for the lower surfaces. The aircraft letter was in the I. Gruppe colour of white.

Originating from a 1935 design intended to produce a replacement for the He 60, the He 114 had finally entered production in 1938, by which time it was obsolescent. Contrary to its original design concept, the small numbers produced were used primarily for coastal reconnaissance work with the Küstenfliegergruppen. A few were operated from Kriegsmarine warships by 5./BFGr 196. This He 114 A-2, 7R+EH, of 1./SAGr.125 is representative of the main production variant. The Werknummer 2562 was marked in very small digits near the rear edge of the fin in typical Heinkel company standard style. The Staffel colour of white was used on the spinner tip. Camouflage was 72/73/65 with yellow added as a band around the rear of the fuselage and beneath each wing tip, but not under the engine cowling. On the Eastern Front the type was operated in the Black Sea area but was eventually phased out in favour of the Ar 196.

This aerial view of some of 2./ZG 76's Bf 110s also captures an interesting element. Most units used obsolescent aircraft for communications or just general hack machine duties. The Do 17, 7A+AM, was from the 4.(H)/Aufklärungsgruppe 121, its pre-war 61/62/63 camouflage scheme on upper surfaces identifying it as an old machine. The Balkenkreuz on the fuselage has been applied in the prevailing style, although the ones on the wings appear to have remained in their extreme outboard position. The problem of the old centrally positioned Hakenkreuz was resolved in a very distinctive manner by applying a huge Hakenkreuz over the entire fin and rudder, which was first painted out with RLM 61. Many units utilised obsolete or obsolescent aircraft as general hacks for logistic support. Whether Aufkl.Gr. 121 was sharing facilities at the same base as the Bf 110s of 2./ZG 76 or was simply passing through the base at the time, is unknown. The Bf 110s, in full 70/71/65 camouflage dates the photograph to the early part of 1940. The Balkenkreuze on the two machines visible were of incorrect form. They should have had the older proportions, with narrow white edging.
(J. Vasco)

France, 1940 and a most unusual example of standard codes. This Ju 88 of 5.(F)/Aufkl.Gr.122, had its code F6+TN marked in a style similar to that seen on some German aircraft in WWI. On the original print the black 'F6' and 'N' of the code could be seen applied in the same ornate manner as the red 'T', outlined in white. The unit's earlier aircraft, Do 17 Ps, are known to have used the same style of lettering, so the 'marking tradition' of this unit extended back to at least 1939.

This photograph provides a closer view of the code style, this time on an He 111 H, F6+RL, 'Yellow L' of 3. Staffel. This flamboyant form of unit code appears to have been unique. No other unit so far photographically recorded, exhibited such a style. However, this very individualistic manner of marking appears to have been phased out around 1940. (D. Wadman)

The Do17 P variant had succeeded the Do 17 F in the photographic reconnaissance role but, other than a change of engine type, there was nothing significant to distinguish the two versions. The machine in the foreground belonged to 1.(F)/Aufklärungsgruppe 120 which, judging by the chipping of the paint on the edges of the propeller blades and the white paint of the spinner, had already seen considerable flying time when photographed. The 1. Staffel badge, a pair of 'reindeer antlers and skull', was usually marked in black on a white shield, but this example was of more subdued form and colouring. The aircraft in the background presents a mystery as it appears to have the code F6+KH, the white 'K' and last letter 'H' indicating a 1. Staffel/I. Gruppe machine, like the one in the foreground. The first two characters of the code group are indistinct, but the second is definitely a '6'; and that restricts the code group possibility for a reconnaissance unit to just 'F6' (Aufkl.Gr.122). The easily identified emblem on the cowling of the aircraft in the foreground also makes identification of that machine's parent unit equally unambiguous. Possibly 'KH' had been transferred to Aufkl.Gr.120. Depending upon pressure of duties and losses, such transfers between units were not uncommon when the fighting was intense.

The Russian campaign absorbed the great bulk of Luftwaffe operations and photo-reconnaissance duties were critical over the vast areas. This Ju 88 D, of Aufkl.Gr.122, F6+D?, had had its standard 70/71 camouflaged upper and side surfaces overpainted with temporary white, but forward of the wings the engine cowlings had been left in their original colouring. The spinners were painted red, and the aircraft letter 'D' in the code group appears also to be red. Yellow tactical colouring had been added in the form of a band aft of the code, and on the lower half of the front section of each engine cowling – a somewhat reduced area of application. Yellow was also absent from beneath each wing tip.

A tense moment as an engine fire erupts in the port engine of Q1+VB, 'White V', a Bf 110 F-2 of Stab./NAGr 1 serving on the Eastern Front. Only the last two letters of the code are visible (VB). The large, skeletal hand emblem on the nose is unidentified, possibly a personal one. The spinners were two-thirds black/one-third white with yellow tips, indicating that it was from the Stab of III. Gruppe. The undersides of the wing tips were yellow, and a yellow band was marked around the rear of the fuselage. Camouflage was quite dense, probably a 74/75 mixture of mottling over the main uppersurface pattern, with 76 lower surface colouring.

Another reconnaissance type to endure the savage Russian winters was the Bf 110. This Bf 110 E-3, 8H+?? of I./Aufkl.Gr.(F) 33, had had its 74/75 camouflage overpainted with temporary white. However, unlike the Ju 88 D seen opposite page top, forward of the wing the yellow tactical colouring had been extended aft to a point approximately two-thirds of the way back on each engine. Note also that the white temporary finish has been confined to exactly the same area on the upper portion of each engine cowling, ending in a crisp vertical line, indicating that the yellow tactical colouring had been painted right around each engine cowling. As on the Ju 88 D, forward of the wing the remainder of each engine had been left in its original camouflage colouring. The yellow fuselage band can be clearly seen, but the yellow on the wing tip of the port wing is only just discernible where it curved around the leading edge. This more extensive suite of tactical colouring was used for both the Balkans and the earliest period of the Russian campaign, which gives some idea of the time period. The glass panelling in the belly of the fuselage for the Rb50/30 camera installation sited aft of the pilot's position can just be seen. A pair of ETC 50 racks were fitted to each wing.

A reconnaissance Do 215 B-1 (originally issued to 3.Aufkl.St./Ob.d.L in early 1940), with the type of camouflage modification seen on some reconnaissance Do 17s during the air fighting over Britain in 1940 (see Volume Two, page 343). In this instance, soft-edged lines of 71 have been used on the 65-coloured undersurfaces. This machine was photographed in August 1942 when it was being handed over to the Royal Hungarian Air Force at Shitomir. (E. Sommer via D. Vincent)

Ar 196 A floatplanes served with the ship-borne Bordfliegergruppen, having been issued to the capital ships of the Kriegsmarine in 1939 for sea reconnaissance work. The type was also used extensively for coastal patrol and reconnaissance work. It underwent modification over time and several variants were produced, close to 600 airframes eventually being manufactured, a testimony to the value of this type. Camouflage was 72/73/65, as seen here on 6W+NN of 5./BfGr 196. Except for the two that could be seen from above, the float struts were painted in 65-colouring, forward of the wing leading edge. Rubber nose bumpers of floats were sometimes painted black, or in one of the two camouflage greens, but on this machine they appear to be red, the colour used pre-war on floats. Each step and hand-hold flap has been outlined in white, also a common practice on this type. The pale chord-wise stripe near the wing tip was a canvas cover strip, the fresher paint appearing lighter in tone. The finish overall was immaculate and the propeller blades show little sign of wear, so this was probably a relatively new machine when photographed.

An oddity in terms of markings is seen here, especially for the Ar 196, which almost always conformed precisely to regulations in such matters. The aircraft was coded K3+CO, which would indicate that the parent unit was Seenotstaffel 6. However, that Staffel served almost exclusively in the Mediterranean area. Given the Ar 196's conformity to markings regulations, one would have expected a white band marked around the rear end of the fuselage. It has been suggested that the large white number '4' on the fin denoted ownership by the 4. Staffel of Seenotbereichkommando, based at Utrecht in Holland for operations over the North Sea and English Channel. The codes used by Ar 196-equipped units often cause confusion and it may be that this machine was transferred without a change to its existing codes – which could possibly explain the identity denoted by the '4'. Careful examination of the original photograph shows no evidence of a white band having been painted out, indicating that this machine had perhaps been transferred before being sent to the Seenotstaffel 6 operational area.

Fw 200s bore the brunt of long-range armed maritime reconnaissance duties during the early part of the war, making valuable contributions to the U-boat offensive by shadowing convoys, as well as carrying out anti-shipping bombing attacks and mine-laying. The RLM's official camouflage directive of 2 August 1941 for the C-1, -2, -3 and C-3/U-4 variants stipulated colours 72/73/65, as by then the type was being used exclusively for maritime operations. The boost to defensive armament introduced with the Fw 200 C-3 can be seen here - a hydraulically-operated dorsal turret armed with a MG 151 cannon and a MG 151/20 cannon in the ventral position. The last two letters of the code group (BS) can be discerned, with the 'B' thinly outlined in the 8. Staffel colour of red, and just the spinner tips in the III. Gruppe colour of yellow. This was most probably an 8./KG 40 machine, photographed at its French base.

RIGHT: The improved Hs 126 B-1 variant had started to reach the Luftwaffe by mid-1939. Little distinguished it from the older version externally, but engine power had been increased and the wireless equipment upgraded. This example, belonging to Aufkl.Gr.(H) 23, had standard 70/71/65 camouflage. The camera-aiming lines were a little different on this aircraft, in so far as the upper (shallow angle) line was marked with a broken white line. The rest were marked as usual with the next line in red, then one in yellow and the last, white. Each line was marked with a number, starting at '50' for the upper (broken) line, then 100, 500 and 1000. As with most Hs 126s used operationally, the wheel spats have been removed to eliminate problems associated with operating from primitive airfields. The fixture attached below the base of the wing strut was a temporary bomb rack, held in place with two short horizontal struts and one long vertical one. All three were painted in RLM 65.

This Hs 126 B-1, H1+MK of 2./Aufkl.Gr.(H)12 has also been overpainted in a temporary white finish, but in this instance the yellow band (approximately 900 mm wide) has been left free of any covering. The yellow around the front edge of the engine cowling had also been left uncovered and in this view it can be seen that no yellow has been used beneath the engine cowling as in the previous photograph. However, unlike that aircraft, the white angles of the Balkenkreuz on the fuselage have been left untouched and emphasised by the narrow edging of the original 70/71 background colouring that had been left free of the white paint. The Hakenkreuz has also received similar treatment.

It was not until the early summer of 1942 that the first of the Fw 189 A-1 and A-2 aircraft reached the war front in numbers. This was no reflection on this very successful design, but rather RLM perception that the Hs 126 was already doing an adequate job in that reconnaissance role. In the end, the Fw 189 would equip a large number of Luftwaffe units and undergo steady refinement, serving until the last days of the war. This example, an Fw 189 A-2 of 5./Auf.Kl.Gr(H)12 was photographed in Russia, the yellow tactical band just visible on the port boom. The wing tip undersides were probably marked in yellow, but it was rare to see an Fw 189 with yellow beneath the engine cowlings. Camouflage was standard 70/71/65. The 5. Staffel badge, showing an Arab boy in a Fez, also appeared in a variation on the 11. Staffel emblem. There may be a link with former personnel who had served with 4. Staffel, the only unit ever to operate the type under control of Fliegerführer Afrika. All the other known badges of this unit's Staffeln have no relationship to this North African influence in their design.

RIGHT: Abandoned during the German retreat in the Libyan desert, Hs 126 A-1, 5F+JK, of 2./Aufkl.Gr.(H)14 wears a Maintenance Unit modified scheme of 79-colouring applied as a dense mottle over the original 70/71/65 colouring, including the spinner, leaving only the tip in the I. Gruppe colour of white. The application was not so densely applied on the engine cowling ring and cooling gills, where the original 70 green is more easily seen. A white tactical band had been added to the rear of the fuselage. The fuel drum beside the aircraft has its end painted in regulation identification 04 yellow, with details in black. (D. Cormack)

I./Aufkl.Gr(F)14 carried out long-range operations using the Bf 110. This particular Bf 110 F-3, 5F+YK, was camouflaged in 78/79 to which RLM 80 has been added, post-production, the colour being issued only to field units (i.e, maintenance units and operational units). The ribbon effect was a style seen also on some other aircraft, amongst them a Ju 52 and a DFS 230, possibly all of them having passed through the same maintenance unit, a number of such facilities having been made available to the Luftwaffe in Sicily and Italy. A white tactical band had been added to the fuselage, the only such addition usually found on aircraft serving in North Africa or the Mediterranean areas. The spinners were halved in white and red, respectively the Gruppe and Staffel colours, as no Staffel colouring was used on the unit code. (J. Vasco)

Often forgotten when the reconnaissance role is mentioned, was the daily weather. Reconnaissance duties were tirelessly carried out in good weather and bad so that operations could be planned and executed. This Ju 88 A, D7+LH, was serving with Wetterkundungstaffel 1./Ob.d.L from Bad Zwischenahn at the time, having just landed from that unit's 1,000th operation. Its camouflage of 70/71/65 had been given additional sweeping strokes of 02 to break up its dark colouring, blurring the edges of the white angles of the fuselage Balkenkreuz in the process. All four characters of the code have been applied in black. One of the propeller spinners was a replacement, the nearest having a dark (red?) tip to its 70 colouring, while the other appears to have a white ring over a colour paler than 70. Given that both Gruppe and Staffel colours were white, it would seem likely that the port engine had the replacement spinner.

The Ju 290 design had come to fruition with the roll-out of the pre-production Ju 290 A-0 series of two machines, followed immediately by five A-1 machines, all of which were immediately taken up by the Luftwaffe to plug the gap in its urgent need for large capacity transport aircraft. A change of role then occurred when the Fliegerführer Atlantik raised the need for a long-range capacity type to track Allied shipping in the mid-Atlantic. The A-2, A-3 and A-4 variants filled the role, each introducing improvements for the reconnaissance role – longer range and better armament. Next to emerge from the process was the Ju 290 A-5 variant, only eleven being built (Werknummern 0170 to 0180), production commencing at the end of 1943. KR+LK, seen here in the winter of 1943/1944, was the last of the batch, the last two numerals of its Werknummer – '80' – stencilled in black just below the cockpit rim. The maritime environment called for 72/73/65 camouflage and the darker uppersurface colour was restricted to just slightly more than the plan view, leaving the 65-coloured side surfaces to blend with the grey skies of the Atlantic, thus helping to break up the bulky side view. (By that time RAF Coastal Command had made a similar shift in its camouflage system for its Sunderlands, an aircraft of comparable dimensions.) The tracks in the snow show that the Ju 290 had been towed out backwards to the hardstanding. In the background is a Ju 88 with both engines running.

Nahaufklärungsgruppe Bromberg operated Bf 109 Gs, utilising an unusual code structure combining the Kampfgeschwadern four-element type plus a numeral as used by Jagdgeschwadern. While of poor quality, this unit was so rare that any detail is useful. Coded N5+DK in black, this Bf 109 G-8, photographed at Bayreuth-Bindlach post-war, also was marked with a large red '4', thinly outlined in white, just behind the cockpit. The number, in the same colours, also was repeated in miniature on the leading edge of each wing. Use of red for the number accorded with Kampfgeschwader 2. Staffel colouring. Camouflage is difficult to discern, but appears to have been the simplified form used by most high altitude reconnaissance aircraft; 77 on upper and side surfaces, with 76 for the lower surfaces (see Volume Two, page 344). This helps to untangle the unusual camouflage seen on this and the other aircraft shown in the next photograph. The extreme pale colouring, while useful at altitude, was a distinct handicap late in the war when the aircraft was on the ground. To this extent, 75-colouring appears to have been added to break up the starkness and to offer some protection. On the port side, 75-colouring has been extended onto the fin, emphasising the Hakenkreuz. (A small number '2', possibly black, was marked just above the elevator, but its significance is unclear.) However, on the starboard side, the rear section of the fuselage appears to have been left in its original pale colouring – the strong sunlight has produced a washed-out appearance but does not account for all of the pale appearance, and a distinct change can be seen aft of the diagonal dividing line running through the Balkenkreuz, the 74-coloured centre of which is absent in the upper left hand portion. The white angles of the marking however are present, part of the original high altitude camouflage scheme, which coincided with the angular downward extension of a 74-coloured segment. In contrast to the rear portion of the fuselage, from the Balkenkreuz forward, a very soft mottle of 75 appears to have been used to tone down the fuselage side surface colouration, extending to the bottom of the fuselage and to the engine exhaust line. A wide band of the original pale colouring can be seen where the engine cowling meets the fuselage, close to the cockpit. (J. Crow)

Another Bf 109 G-8, coded N5+EK 5, from the same unit, provides a more extensive variation on the radically modified original high altitude camouflage of 77/76. In this instance it appears to have had the lighter of two uppersurface colours extended right down to the bottom of the fuselage sides, and right down to the exhaust line on the engine cowling, and then had very small, random patches of a darker colour added in just a few places, which was broken up with thin meandering lines of the original 77 surface colour. A distinct change in colour can just be detected along the leading edge of the wing where the original 77 uppersurface colour wraps around to meet the 76-colouring along a line just in front of the wheel-well. The number '5' is repeated on the leading edge of each wing, in red with a thin white outline. However, enlargement of the number '5' on the fuselage shows that it was just a white outline over the existing camouflage, the background giving the illusion of a darker colour (distinct, discrete parts of the added camouflage colours can be seen when significantly enlarged). As on the previous aircraft, the original white outline form of Balkenkreuz has been retained, filled in with the camouflage colours. The spinner spiral is correct for the late war period in the West. (J. Crow)

LEFT: Close up of the nose of Ju 290 A-5, 9V+BH, of 1./FaG. 5 showing the soft-sprayed edge used for the 72/73-scheme where it met the 65-coloured side surfaces. Upper surface camouflage on these large aircraft was restricted to the plan view. Like the BV 138, this late addition to the maritime reconnaissance role, was never available in large numbers but did valuable long-range work.

The ubiquitous Fw 189 served right to the last days of the war. This example, 5H+RK W.Nr. 125215, of NAGr.16, camouflaged in the older colours of 70/71/65, wore yellow on rudders, wing tips and around the engine cowlings – an obsolete mixture of tactical markings given the end of war date of the photograph. The aircraft letter was thinly outlined in the I. Gruppe colour of white. The area behind the '5H' portion of the code had been painted out, indicating that the aircraft code had been re-applied in smaller size, in accord with the 1943 RLM directive: if so, then this particular aircraft had seen long service, which would explain the older camouflage colours. Alternatively, the aircraft may have been transferred from another unit, although the lack of change to the rest of the markings seems to rule that out. (J. Crow)

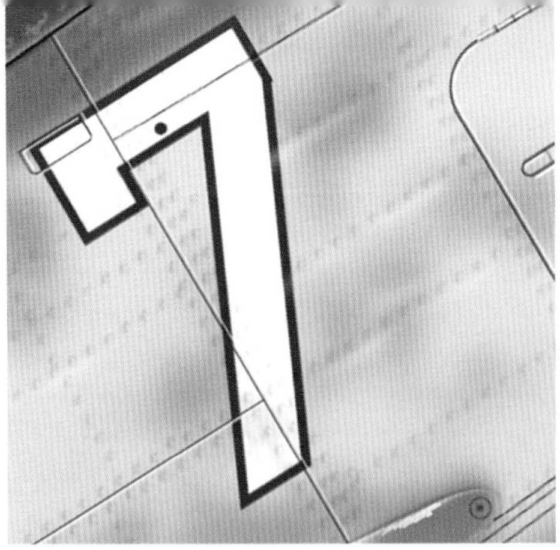

Civil, Commercial and Military Aircraft

He 70 W.Nr. 403, was the third aircraft to receive the civil code D-3. The other aircraft registered with this code were a Sablatnig P 1 and a Junkers Ju A 50 ce. He 70 D-3 eventually bore the registration D-2537 when it entered service with Luft Hansa (see Volume Two, pages 352–353). These views show that the registration also appeared on the upper surface of each wing as well as on the undersides. Note also that the shortened walkway on the port wing may be compared to that on the starboard wing. The German national tricolour marking of black, white and red horizontal stripes had not yet been applied to the vertical tail surfaces when these photographs were taken, indicating that they were probably taken soon after completion at the Heinkel factory and at the beginning of the air-testing programme. The Heinkel badge appears on the fin on both sides with the Luft Hansa logo on the rudder only on the port side.

When the first He 70 entered commercial service with DLH in March 1933, it operated at speeds which matched those of contemporary RAF fighters, a remarkable performance which enabled it to establish several commercial speed records. D-UMIM displays the type's superb streamlining and the DLH colour scheme of pale grey with black trim. The lightning flash marked across the dark cowl, using the pale grey base colouring, was adopted for the machine to enhance its type name of 'Blitz'. The company name 'LUFTHANSA' appeared in white block letters below the first horizontal stroke of the 'lightning' line. Bird names were used for naming the He 70 fleet, with this aircraft's name of 'ALBATROS' marked in similar style and colour below the next horizontal stroke; other known ones were SCHWALBE (Swallow) and SPERBER (Sparrow-hawk). The national tricolour marking, as shown here, had been extended to span both fin and rudder on production aircraft (see Volume Two, page 353). (Lufthansa)

This close-up view of the cabin area of D-UMAS provides some idea of how slim was the fuselage, holding just four passengers. The open cockpit door behind the ground staff handler had some of the instructions for the second pilot/radio operator attached to its inner face. The elliptical shape of the wing and its broad wing walkway marking are evident. The inboard corner of the landing flap was edged with a black (rather than red) line and had 'nicht betreten' (Do not walk) marked on it in the same colour. This machine bore the company name and the individual name 'HABICHT' (Hawk) on the engine cowling. (Lufthansa).

The Ju 52 started its career with Deutsche Lufthansa in 1932. D-AVUP, KURT STEIDEL, shows the broad black panel that stretched back over both upper and lower surfaces of each wing, immediately behind each engine, extending to the flaps. The black undercarriage struts and stylish trim to the wheel spats were other features of the type, as were the black engine cowlings and sweeping trim. The wing dimensions allowed the marking of the civil registration in a straight line. (Lufthansa)

The Ju 52 had several configurations available for its interior, which could seat up to 19 passengers depending on layout used. Compare the relative austerity of this layout with the two following photographs. (Lufthansa)

In this aft-facing view, the interior can be seen to be more elegant, with a chrome-edged parcel rack replacing the earlier netting form, flush panelling and lighting, and more comfortable seats. Passenger entry was via the centrally placed door in the rear bulkhead, with a separate door marked 'W.C.' for the toilet, on the right of the view. Once the type entered military service, such elegance was replaced by a variety of basic bench and other forms of seating and stretcher fittings, depending on the role of the particular machine. (Lufthansa)

This forward-facing view is of a more luxurious layout, with each seat finished with leatherette and a built-in headrest. The single central control column, with throw-over arm for the two pilots, identifies this as a relatively early version of the Ju 52, two separate control columns becoming the norm soon after production began. (Lufthansa)

Another type soon destined for a military career was the He 111. The He 111 C-02, D-AQYF named 'LEIPZIG' illustrates that names were marked on both sides of the fuselage, revealing also the curve in the application of the registration on the wings, required to accommodate the semi-elliptical wing shape. Note also the wing walk area inboard of the engine, marked with a black outline.

In the spring of 1938 the Ha 139 B series prototype, the V-3, W.Nr. 217, registered D-ASTA, joined the others in service with Luft Hansa. It carried the usual pale grey finish with yellow safety markings. However, the modified tail fins saw the Reichs-und Nationalflagge confined to the rudders. In the view below the aircraft name of 'NORDSTERN' has not yet been added. It was eventually taken over by the Luftwaffe, repainted in 72/73/65 and, initially, designated as Ha 139 V3/U1. It was used by Küstenfliegergruppe in a reconnaissance role, before being further modified and designated Ha 139 B/MS, marked P5+GH, and employed for mine-sweeping duties. After being modified to the same standard as the Ha 139 B, the other two former Lufthansa aircraft were also employed by the Luftwaffe, seeing service with K.G.z.b.V. See in the logistics support role, transporting ammunition and supplies for the ground forces during the invasion of Norway. (Lufthansa)

LEFT: The first unpainted 'Condor', the Fw 200 V1 W.Nr. 2000, is having its engine tested outside the construction hangar at the Focke-Wulf factory in Bremen around the middle of July 1937. At this time the machine was still finished in its primer colours before receiving its first colouring and registration markings of D-AERE.

BOTTOM: After extensive flight trials it was decided for propaganda purposes to try to make the first airliner Atlantic crossing non stop from Berlin to New York and the Fw 200 V1 was chosen for the task. The machine was taken back to the Focke-Wulf factory where it underwent extensive conversion, especially to the wings, and the addition of several fuel tanks fitted in the fuselage containing around 10,000 litres. The machine eventually re-appeared with a new registration of D-ACON but was still named the 'BRANDENBURG'. The flight commenced on 10 August 1938 from Berlin Staaken airfield at 20.05 hours and the aircraft eventually landed at Floyd Bennett Field, New York after a flight of 24 hours 36 minutes. The flight proved that future passenger-carrying transatlantic flights were feasible.

LEFT: The colour scheme of Fw 200 V 1, D-AERE (initially named 'SAARLAND') was a little unusual, starting out with the DLH pale grey overall finish (including the spinners), but with the black trim not applied to the engine cowlings or the nose, as would occur later and also on its contemporaries. The name SAARLAND was eventually transferred to one of the later prototypes, and the V-1 renamed 'BRANDENBERG', but reversed the Company practice of having the registration on the fuselage marked large, with the given name in smaller letters. The company logo was marked larger on this type. (Musal via Santos)

ABOVE: In this view of Fw 200 V2 D-AETA named 'WESTFALEN', the extent of the black paintwork to the nose section can be seen. Note the Focke-Wulf company logo and names behind the cockpit, in the pale grey finish, with the Luft Hansa name and logo on the nose section. The engine cowlings were also painted in the company pale grey finish. (Lufthansa)

Photographed in Spain, at Leon, this Ju 52 was serving with the Luftfeldpoststaffel whose task it was to deliver mail to all military locations (hence the emblem on the nose – an eagle clutching a letter in its beak). Colouring was unusual, clearly a darker shade than the 02 of the engine firewall, but much lighter in tone than the black of the usual Junkers' trim. It may have been 62-green as this was an operational area, and that colour was being used by some of the front line German types. The under-carriage, engines and inboard section of the wings were finished in black trim, indicating that it had previously seen civilian service. (D. Vincent)

This example of an air sea rescue He 59 B carried an interesting variation on the revised markings applied to military second-line aircraft, to replace what were technically 'civil registrations'. The revised 'WL' (for Wehrmacht Luft) form of marking was introduced on 1 January 1939, replacing the existing D-series civil registrations, converting what would have been the D-AHAN registration to WL-AHAN. This was done to distinguish military second-line aircraft from civil aircraft. The WL form of marking caused most aircraft to have their existing 'D' registration cramped to accommodate the revised prefix letters, but this machine had obviously been freshly painted before the markings were added. The Red Cross marking was thinly edged in what appears to be red rather than black, making it stand out on the 02-overall finish. The use of 02 for the overall finish was somewhat unusual as ambulance aircraft of any type were at that time traditionally painted in white overall. He 59s that served with the Seenotflugkommando on air-sea-rescue duties only abandoned their all-white finish after the infamous July 1940 debacle which resulted in an order by the Allies to shoot down such aircraft if encountered near any ongoing sea/air operation (see Volume Two, pages 298-299). In the distance an He 115 can just be seen.

The crisis of July 1940 and the subsequent shooting down of some of the white painted He 59s of the Seenotflugkommando units prompted a rapid shift to full camouflage for aircraft, land or sea based, used for such work. Red Cross markings on a white disc, marked in the six positions normally occupied by the Balkenkreuz, were combined with the Stammkennzeichen as shown on this Fw 58, TR+AU, camouflaged in standard 70/71/65, the pattern of which can quite readily be distinguished. The rudder is slightly offset to starboard, catching the sunlight, producing a slightly lighter appearance. The civil Reichs und Nationalflagge marking was replaced by the military Hakenkreuz.

The Luft Dienst (air service) played a significant second-line role, carrying passengers and various items for the Luftwaffe. The unit used a wide variety of types, many of them obsolete aircraft as seen here on BA+DX, a Do 17 (probably an F-1 rather than an E-1 given the duties of the Luft Dienst) which belly-landed at Prague/Rusin in 1940. It had retained its 1938-style camouflage of 61/62/63/65, there being no requirement for specific colouring or camouflage for its second-line role. The retention of the white colouring for the rear section of the spinner is unexplained by the context of the photograph. Either this was a reserve machine, still retaining its Stammkennzeichen, or it had had any unit markings very carefully painted out and a Stammkennzeichen substituted, because that system was not introduced until 24 October 1939 (see Volume Two, page 362). The former option appears the most likely, in which case the white spinner band would most probably relate to its Luft Dienst service role.

The construction of extremely large capacity cargo gliders had grown from the failure of the intended invasion of Britain, which had highlighted the shortcomings of having the means to rapidly transport logistical material to the battle front, something which had to be addressed for the forthcoming invasion of Russia. The urgency may be judged by the fact that a specification had been issued to both the Messerschmitt and Junkers companies within two weeks of the postponement of Operation Seelöwe – with a requirement for studies to be submitted within 14 days. The same relentless pace continued during the design and construction phases, with the first prototype being ready for pre-flight testing by February 1941.

In June Grossraumlastenseglergruppe Me 321 was formed, W2+SZ being one of its Me 321 A-1 machines. Camouflage was 70/71/65 and codes were marked in black. Initially the unit comprised three Staffeln, each with six aircraft. The code allocation for the type was distinctive, each Staffel being allocated a 'W' plus a consecutive number sequence, i.e. W1, W2, W3. Two code letters, commencing with 'S', completed each sequence, e.g. W2+SP and W2+SZ. Codes W4, W5, W6 and W8, were added when the organization was revised to form four Sonderstaffeln (Grossraumlastensegler), each with five Me 321s, for use on the Eastern Front.

No less than five Me 323 D-1s can be seen in this photograph, all camouflaged in the usual 70/71/65 finish. The location is not known, but none have any tactical markings and the code on the one in the centre was all in black, the last two letters being 'QS', probably part of a Stammkennzeichen rather than a unit code. Some idea of the size of these machines may be gauged by the scale of the personnel working on them.

RIGHT: The powered version of the massive Me 321 glider was the Me 323, this E-2 version operated by I./TG 5 was photographed on the southern sector of the Eastern Front. Low drag EDL 151 turrets were introduced on this model to try to provide some defensive armament, but the sheer low speed and handling characteristics of the type made it easy meat when faced with fighter aircraft. Camouflage on C8+E ? ('Yellow E') was standard 70/71/65 plus the addition of yellow beneath the wing tips and as a band around the rear of the fuselage; the latter in this instance, even at one metre, appeared disproportionately small in width for the size of the type.

A damaged Ju 52 of 2. KGr. z.b.V.106, photographed in Russia during 1941. The unit's Narvik-Schild, adopted after taking part in the Norwegian campaign, is marked on the flat panel forward of the cockpit. The anchor marking was painted in yellow, the same colour used for the '2' marked on the cowling ring of the central engine. With the exception of the outboard upper quarter of the two outer engines, the cowlings were painted in RLM 04. The camouflaged quarter retained had had a broad yellow band, marked chord-wise across it (a form of marking also recorded, later in the war, on Ju 52s of 2. KGr.z.b.V 700 when part of Transportgeschwader 4). The yellow on the central engine had been restricted to the lower half. Presumably the stripe was a form of visual reference for formation flying. (A similar form of marking had been recorded on Ju 52s used for the invasion of the Low Countries in 1940, white chord-wise stripes being recorded for one machine, but on all three engines in that instance.) A yellow band around the rear part of the fuselage was probably present. Camouflage was 70/72/65, as on the Fw 58, TQ+AZ, in the background marked with a yellow band behind its Balkenkreuz. (K. Kössler via Jessen)

By the winter of 1942-43, composition of the transport units had undergone revision, temporary combinations of individual units being made into Transportgeschwadern for larger operations. The aircraft's individual and Staffel letters were painted in smaller characters at the tip of the fin, as shown here, 1Z+LX of IV./KG.z.b.V.1., and a secondary code was added to identify the temporary Transportgeschwader (see Volume Two, pages 297-298). An overall coat of temporary winter white camouflage has obscured the unit code of 1Z, leaving the last two characters – the 'L' in white roughly outlined in the greens of the 70/71 camouflage, and the 'X' in black, superimposed on the yellow band. The Balkenkreuz and Hakenkreuz have also been left narrowly edged in the background camouflage colouring. (via M. Jessen)

The history of development of the Ju 252 was a chequered one, commencing as a design study for a 21-passenger aircraft for Luft Hansa in 1938, and ending as a heavy transport for the RLM. Work on three prototypes commenced at the Junkers Dessau plant in July 1940 and the V1 flew in October 1941. Two other prototypes joined the flight test programme in the winter of 1941/42. The Ju 252 V4, DF+BP, served as the production prototype and began test flights in early Spring 1942. Despite its usefulness, material shortages were already cutting into aircraft production programmes, and as a result only 15 were ultimately produced. Despite the production designation of Ju 252 A-1 being adopted, all the airframes were allocated a Versuchs number, from V1 to V15. The V4 was pressed into Luftwaffe service, ferrying engines to Bordeaux-Mérignac in support of the He 177 evaluation by KG 40. The remaining aircraft, V5 to V15, were delivered to the Luftwaffe during the last part of 1942. The V5, DF+BQ, seen here in service, was assigned to the Viermotorigen Transportstaffel when it formed at Tempelhof on 2 January 1943. The unit was soon renamed Lufttransportstaffel 290 and a Ju 290 can be seen in the distance, which points to this photograph having been taken at that time. Note that, like the DF+BQ (opposite page top), the aircraft had a metre-wide yellow tactical band around the rear of the fuselage. There is no evidence of yellow having been applied beneath each engine cowling, and unfortunately the wing tip area is not visible in this view, although it is most probable that yellow has been added. The huge Trapoklappe is in the open position for loading. Camouflage was 70/71/65 and the aircraft appear to have retained their Stammkennzeichen rather than using a code group, at least in the transport role.

This example was probably the Ju 252 V 11, which had no EDL 131 turret fitted aft of the flight deck. The location appears to be Tatoi, in Greece, as there are other photographs showing DF+BW at that location in August 1943 during a transport flight. Again a yellow band has been applied around the rear of the fuselage and in this view yellow has been used beneath the wing tips.

A rare bird – He 111 H-2, D-ACLO, of the Axis Armistice Commission, which was captured at Maison Blanche, Algeria, in 1942. The diplomatic nature of its operator required deletion of all military markings, (including the military version of the Hakenkreuz), and use of full civil markings, with the National Flag emblem across fin and rudder and civil registration, in black. However, the aircraft retained full 70/71/65 camouflage, albeit with a division between upper and lower surface colours along an unusual line for the type. (J. Crow)

Abandoned at Fuka, this Go 242 B-1, DQ+??, wore a modified camouflage scheme created by the original 70/71 upper-surfaces being oversprayed with 79 through a net used over parts of the airframe to produce a simple 'mottle' effect. The thin 79-colouring has been eroded along parts of the leading edge of the wings, revealing the original 70/71 camouflage in places. The 70-colouring had been retained around the cockpit and the original mottling in 70 and 71 on the forward fuselage is still just visible as a ragged lower edge to the 79-colouring.

This is not the same Go 242 B-1 shown in the photograph at the bottom of page 115, but it is from the same production batch. DQ+AX had what appears to be a tropical finish of 78/79, but applied with greater care, most probably at a maintenance depot, since this aircraft was one of several which had been redeployed to a theatre of war that hastily required a distinctly different form of camouflage (see following photographs). The lowered flap shows signs of overpainting of a darker colour (the original 70/71), more clearly seen along the trailing edge area, especially near the left-hand corner. Note the absence of a Balkenkreuz between the letters of the Stammkennzeichen – that marking usually being confined to the rear, removable portion of the box-like fuselage – a horizontal black stroke being used instead. (J. Crow)

A somewhat grainy, poorer photographic image of DQ-AX, showing the yellow tactical colouring beneath the wing tip and as a band around the rear section of the fuselage. The Balkenkreuz had been taken as a souvenir by Allied troops. The poorer quality of the image helps to illustrate how this factor can influence perceptions about camouflage. The location appears to be North Africa, possibly Fuka, where others of this kind had been abandoned in the retreat. The presence of yellow tactical markings appropriate to the Eastern Front was a factor encountered not infrequently with the rapid re-deployment of some units and types to different battlefronts at short notice and for anticipated short duration. Desert colours were also not inappropriate to the Russian deployment in summer, when the landscape on the open Steppe was a uniform golden brown. (J. Crow)

Part of a group of Go 242 As photographed at Stendal. NE+ZZ, shows the 70/71 application to the uppersurfaces of the fuselage and also has 04-yellow tactical markings. When the type first entered service in 1941, six Staffeln were formed as part of Go 242 Staffeln der Luftwaffe, and used for home and rear area supply duties. As such, they retained their Stammkennzeichen but had tactical markings added where appropriate. (H. Jonas via D. Vincent)

These Go 242s of an unidentified Lastensegler unit were all heavily camouflaged with two dark colours on the top and upper side surfaces and were apparently abandoned with other aircraft (a late model Bf 109 G, with a camouflage net draped over the rear of the fuselage and a Fw 190 can be seen in the background). The lower side surfaces of the fuselage to the left had the upper camouflage colours applied in broad bands that angled forward and downwards. These had been extended to the bottom edge of the fuselage as a dense mottle. The underlying pale lower surface colour had bled through in places, slightly lightening the darker colours, which appear to be more like 82 and 83 than either 81 and 82, or 70 and 71. (The distinction is more clearly visible – under magnification – on the nose of the aircraft immediately behind.) On the nose beneath the cockpit, '018' has been marked in white. A red, horizontal dash and number '- 24' had been marked on the rear section of the fuselage, the first part of the code, which would have been applied in small characters, not being visible. The aircraft in the foreground wears the same colours, the deep shadow thrown by the wings making it appear that a darker camouflage had been used. The apparent dark colour on the boom was, in fact, a tarpaulin, but the lowered flap beside it confirms that the two uppersurface colours were the same as those used on aircraft '018'. The transition line with the pale, lower surface colouring (probably 76 rather than 65) has been applied with a soft, ragged edge, as was usual on this type of glider. The '005' number again was in white. There appear to be no yellow tactical marking colours, suggesting that this unit was home based. A red coloured '-12' number can be seen forward of the Balkenkreuz on the fuselage which, like the Hakenkreuz, was in outline form, and therefore correct for the late war period. The overall finish of these gliders indicates that they were either late production airframes or had been repainted at a maintenance facility. The parts spread in front of '005' tell more about the location: two front section engine cowlings for an Me 262, in 02 finish; some wing parts from another type; a glider wing section, still marked WL-VII, indicating that it was an old part; a canopy frame; a fuel tank and other broken parts. These are indicative of a Western European location rather than further east. (J. Crow)

The giant Me 321 glider was given standard 70/71/65 camouflage, its massive bulk making it difficult to disguise on the ground. Note the relative roughness of the dividing line between upper and lower surface colouring along the wing leading edge, the slight sag of the wing covering between the ribs making the surface uneven.

The Ju 290 V1 (alias the Ju 90 B) BD+TX was pressed into service with the Luftwaffe as a heavy transport. It wears standard 70/71/65 finish. The type identification 290 V1 is marked in white on the nose section. It was eventually lost taking-off from an airfield near Stalingrad on 13 January 1943.

Some unique marking situations occurred, such as on this He 111 Z with its twin fuselage (basically two complete aircraft). The two, however, were treated as a single unit, the markings being repeated on both fuselage sections. In this instance the Stammkennzeichen, DG+DY, had been applied in black. The rear section of both fuselages was marked with a yellow 04 tactical band. (D. Wadman)

Full circle. Seen at Prien in Germany in May 1945, D-ASPI, W.Nr. 6710, had been sold to Italy in 1939, and then leased back by Luft Hansa in 1943 for use on the route between Venedig and Riem. During its service in Italy the original engines had been removed and replaced with Italian-manufactured units and fitted with three-bladed propellers. On 21 April 1945, during a scheduled flight between Riem and Tempelhof, being flown by Engelbrecht Tischbiereck, the aircraft was reported as 'missing'. It had landed behind Allied lines – a sensible decision for an unarmed civil aircraft in those last days of carnage. The colour scheme was that of the Luft Hansa fleet, with the Reichs und Nationalflagge and black civil registration marked in the standard positions. (J. Crow)

Ju 52 D-ASPI was not the only transport aircraft found at Prien. On its starboard side was another Ju 52, with camouflage of 70/71/65, of an unidentified Transport unit, which had seen recent service on the Eastern Front with the retreating German forces. Only the final part of the code, +LD, is visible, the 'L' marked in yellow. Yellow also was present beneath the wing tips and underside of each engine cowling, but the yellow band around the rear of the fuselage had been painted out. On the other side of D-ASPI, was KB+ZE in Luftwaffe markings, one of the relatively rare Siebel Fh 104 A aircraft, only 46 of which were built. The colour scheme appears to be that company's pale grey finish as used for the prototypes, complete with the lightning flash-style stripe running from nose to tailplane (see Volume Two, page 395). It appears lighter in tone than the black of the Balkenkreuz, so it is possibly red. Next to it is another Ju 52 fitted with similar type engines and three-bladed propellers as D-ASPI, but wearing a combination of full camouflage and civil markings, the Reichs und Nationalflagge on fin and rudder being just visible; possibly one of the machines seized from the Italian forces. Note also the Fw 189 on the left of the photograph. (J. Crow)

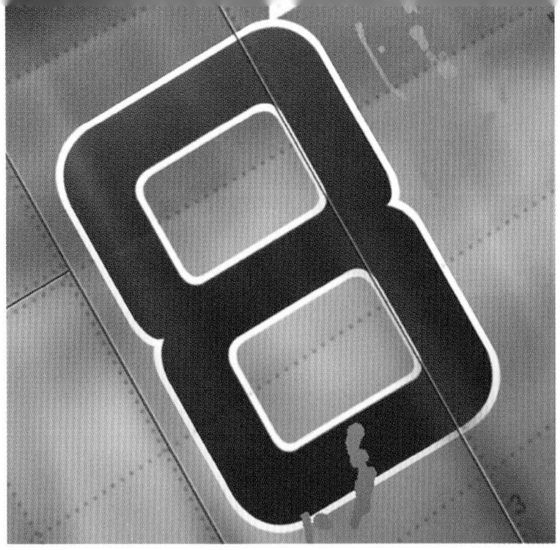

Training, Liaison and Light Aircraft

The Sammelmitteilung of 1 July 1944 contained a directive requiring all training gliders/sailplanes (as well as transport gliders) to be camouflaged – effective immediately – with 81 and 82 on "all parts visible from above and the sides" (see Volume One, page 380). The camouflage drawings issued appear to have not reached all holders of such aircraft and the range of variation which resulted was interesting. On this example of a Grunau Baby II seen post war in Czechoslovak service, the original Stammkennzeichen, LX-U?, is hard to detect against the intense camouflage. The Hakenkreuz had been overpainted with Czechoslovak national markings on the fin and rudder, but otherwise the finish was original. (JaPo)

Even the basic primary glider-type, the SG 38, was required to adopt camouflage of 81 and 82, as this post-war photograph shows. Only the port wing and fin covering remain of the original finish, but they are sufficient to demonstrate that all gliders, of all classes, were subject to camouflage in the closing months of the war. Careful examination shows that the 05-colouring had almost, but not completely, been covered with a thin misting of 82 followed by heavy mottling of 81. The removable plywood nacelle over the basic seat arrangement is still in clear varnish finish. (JaPo)

There were, however, always exceptions, possibly because of lack of access to the required camouflage paints, or just insufficient time and staff to carry out the task – training schools had a large holding of aircraft and keeping them airborne was the first priority. This 05-coloured DFS 108-49 Grunau Baby II B, LN+SS, ex-D-IX-46, was displayed in England at the war's end. Allocated Air Ministry number VT762, that addition had not been made when the photograph was taken in July 1945. The use of a '+' marking to separate the two elements of the code identifies its previous ownership as Luftwaffe rather than NSFK. Allocated to No. 26 ATC Gliding School at Sutton Bank, it served until 22 August 1952 when it was written off after an accident. In the background are an Fi 156 and Bf 110. (D. Vincent)

Not all glider training was confined to basic instruction for novice pilots. LA-HH was one of the Stummel Habichts used for conversion training for the Italian pilots sent to Germany in January 1945 for training on the Me 163 B (see Volume Two, page 382). The camouflage was 81/82 applied in a loose network of lines over the original 05-colouring as required under the final directive for this class of aircraft. The pale 05-colouring of the front section of the wing and the nose cone had been left, presumably to help in breaking up the outline of the aircraft if seen from above. (F. D'Amico and G. Valentini)

Prior to the open declaration of the Luftwaffe in 1935, the role of flying training within Germany had been the only form of publicly organised flying available under the restrictions that had been imposed regarding military flying. However, the Reichswehr-Ministerium had secretly continued its ambitions for a military flying force and several military designs had been tendered to the private sector during the late 1920s and early 1930s. Amongst them were the Kl 25, Ar 65, Ar 66, Fw 44, Fw 56, He 45, He 46, He 50 and He 72. Soon after the National Socialists won power on 30 January 1933, the RLM became the controlling power for the, as yet, clandestine Luftwaffe. (see Volume One, page 13). The Ar 66 originated from a 1931 concept, designed from the outset as a two-seat trainer. This Ar 66 C, D-IZOF, wore an RLM 02 overall finish with its turtledeck painted in a darker colour, probably RAL 5009 blue from the RAL colour system, or RLM 24, (dependent upon the date of the photograph), with a thin edge line of the same colour (see Volume One Colour Cards). The standard weights and loading and operator data, marked in black can be seen below the tailplane. The vertical tail surfaces were marked with the standard red banner/white disc/Hakenkreuz but without knowing if the starboard side was similarly marked, it is not possible to date the photograph. If the old tricolour was present it would be before June 1936; if not, then after that date (see Volume Two, page 434). (M. Jessen)

The clear evidence of the horizontal tricolour of black, white and red (black uppermost) on the starboard side of this Ar 66, D-IXUL, makes dating the photograph more definitive – before June 1936. That, and the relative age of the type, helps to define the overall base colouring as most probably RLM 02, since RLM 63 did not enter the Luftwaffe painting cycle until early 1936. Unless this aircraft had had a very recent total overhaul and repainting, the latter was unlikely to be the colour. The turtledeck may be presumed to be dark blue (either RAL 5009 or its very close approximation of RLM 24), with a thin extra line of the same colour along its border. Confinement of the tricolour to the area of the rudder immediately above the horizontal tail surfaces matched the positioning of the Reichs und Nationalflagge on the other side.

The He 72, a 1933 design, was one of the most important primary training aircraft designs used by the Luftwaffe and NSFK. D-EKYR, an He 72 B-1, was sharing a ceremony with a Kl 26 L, which still retained a D-numeral registration (possibly D-2282), horizontal band marking of blue and yellow across the rudder, with yellow on the wing leading and trailing sections, and elevators, cut longitudinally by dark blue bands. The fixed horizontal tail surfaces, the central portion of each wing, and fuselage, were in a honey coloured-tone, probably varnished wood and fabric - colours used by the old DVS, in the days when it served its role of clandestinely training pilots for the Reichswehr. Interestingly, if the civil register of the period is correct, there was no allocation made for D-2282 – so, was this aircraft colouring and registration something done just for the ceremony? The He 72, in contrast, had an overall finish of a pale colour, possibly a cream colour, with a dark-coloured cheat line from radial engine to the tailplane, the same dark tone also used for the sweeping, diminishing outline around each wing tip (the latter a form of decoration had been used by some aircraft of the old DVS facility based at Schleissheim). This appears to match the dark band colouring on the Klemm. On 25 March 1933 the new National Socialist government had subtly changed the title of the DLV (Deutscher Luftfahrt-Verband – the German Aviation Association founded in 1902) to Deutscher Luftsportverband, thus retaining the existing acronym, DLV, and easing a centralised regulatory system into place. That was the first, major step in bringing all flying training in the new Germany under the Government's control. From that date forth, any one wishing to learn to fly, or continue flying, had to become a member of the new DLV. Bruno Loerzer, of World War I fame, had superseded the old DLV president as head of the new organization, which was clearly a paramilitary one, with uniforms, ranks, etc., and the existing air units of the SA and SS were absorbed into it. On 17 April 1937 the title was again changed – this time to the NSFK (National Socialist Flieger Korps). The presence of NSFK personnel around the far side of the parade, thus marks the date to April 1937 or later, and was probably the ceremony (at Schleissheim?) where the old DLV/NSFK change was formally being celebrated. The presence of the Klemm, still in its old DVS colour scheme and pre-1934 registration markings, appears to be part of the ceremonial link to that early training organization that had clandestinely trained pilots for the future Luftwaffe.

The introduction of the new four-letter form of Stammkennzeichen registration markings, coming as it did close on the heels of the change from the old 'D' series registrations to the 'WL' revision, caused problems of interpretation at many units holding second-line aircraft. The sheer number of training aircraft in use at the A/B Schulen in 1939 probably swamped the capacity of ground staff to cope with the changes needed while keeping up the intense flying training programme since the war had by then begun. This Ar 66 provides a window into the problems and changes: its original "D' civil registration had to be altered first to accommodate the 'WL' marking and a Balkenkreuz, then the entire registration removed and replaced with the new allocation of RT+NX. The evidence of the overpainting where the original registration had been altered, and an overly large Balkenkreuz added (using in this instance, the horizontal dash of the 'D' registration), can be seen as fresh, paler paint. The Hakenkreuz on the rudder is in the correct position, probably due to the fact that there was no fixed fin above the horizontal tailplane on the Ar 66. The Balkenkreuze, both on fuselage and upper wing top surface, were of the pre-October 1939 style. The ground staff have also added the RT+NX code to the upper surface of the wing, something not usual – another problem of interpretation as staff adhered to the older instruction that both the 'D' and 'WL' markings had to be shown on that surface.

The introduction of the new Stammkennzeichen system in 1939 initially included some restrictions on certain combinations, but those seem to have disappeared by 1941 (see Volume Two, page 356). Even so, certain combinations could be eye-catching, as this one seen on a Klemm 35 marked SF+II, the last two letters being 'I', not the number eleven. Overall colouring of the aircraft was probably 02.

RIGHT: Coming second to the Fw 56 in the trials run by the RLM, the Ar 76 had failed to secure a production contract as a front line aircraft, and was relegated to the training role for the A/B Schulen. Given the high performance characteristics involved in those RLM trials, it might have been expected that the Ar 76 A would have then passed to the advance flying schools, (as did the winner, the Fw 156, in due course) rather than the ab initio establishments. This machine, DH+GF was photographed at Kamenz in 1940-41, the home of both FFS A/B 117 and LKS 1. Overall colour might still have been 63 judging by the fact that the Hakenkreuz remained in the mid-fin/rudder position; if it had been overhauled and repainted in 02, the relevant colour after 1938, one would have expected that anomaly to be corrected – especially given the date. The spinner tip appears to be in a lighter colour, but careful comparison with the highlighted area on the curve of the engine cowl shows that it was also in the same colour as the rest of the airframe. The Stammkennzeichen was marked on the underside of the wing as normal.

In 1926 the Junkers Company undertook development of its successful F-13 design, producing two variants, the in-line engine W 33 and the radial engine W 34; the former appeared the following year. Designed for airline service, it was very successful but finally ceased production in 1934. The Luftwaffe made extensive use of the type for training and communications work, and later also for advanced flying training. This example, ?K+AP, was photographed after the introduction on 24 October 1939 of the Stammkennzeichen four-letter system of military aircraft registration. The earlier mid-rudder/fin positioning of the Hakenkreuz emblem had been painted out, leaving a pale area on the fin, so the overall finish was pre-January 1939. This was most probably RLM 63, which had replaced RLM 02 in 1936, and the overhaul time limitations would have ensured that this airframe had been repainted at least once. The rudder was turned slightly to port, making it appear slightly lighter in tone. The last two digits of the Werknummer, '47', were marked in black at the tip. A shield-shaped emblem just aft of the cockpit possibly identified the training unit to which it belonged, but the details cannot be determined.

Problems of accommodating two letters in place of one produced a range of cramped compromise markings on many second line aircraft. That had been solved on this He 46 by reducing the size of the 'WL' marking to fit the width of the space formerly occupied by the 'D' prefix, in the process making the letters much smaller than normal heights set for registrations. The second problem at many units making such changes involved the introduction of the military Balkenkreuz into the registration group. The horizontal dash, between the 'D' and the remaining four letters, was sometimes converted to a Balkenkreuz, which may have been the solution adopted in this instance. The width of the white angles to the black central cross of the Balkenkreuz had been applied in the style pre-dating the 24 October 1939 changes to the marking; not surprising given the age of the particular aircraft type; and that in turn indicates that the overall finish was probably RLM 63, rather than 02.

The changes which took place to registration markings between 1 January and 24 October 1939 also encompassed changes produced by the reshuffle of Geschwader designations and areas of command. Under the old five-character military code system, in place from at least October 1935, advanced flying training schools had been allocated a similar system (A/B Schulen retaining their civil 'D' registrations) – five characters, but with a slightly different meaning to that used by front line units (see Volume One, page 234). Under that system, codes for training schools all commenced with the letter 'S', followed by a number identifying the Luftkreiskommando, a letter to identify the Staffel within the school and then a one or two digit number to identify the individual aircraft. Thus S4+Q22 seen here identified Luftkreiskommando IV, Staffel 'Q' and aircraft number 22. The intriguing thing is that this code adds to the list of known S-code designations (S2, S5, S6, S7 and the enigmatic S13 and S17) which were used. The fact that this code system was employed only for the advanced flying training aspects has meant that relatively few photographs record their use. The other aspect is that since there were usually obsolete former front line aircraft, which thus could be called upon, if needed, when the reoccupation of the Sudetenland temporarily brought the Western world to the brink of war, they were given front line camouflage. This photograph was taken after October 1939, as the Balkenkreuz beneath the wing was hugely oversize, a common occurrence when the revised proportions were initially promulgated, providing a time frame for the continued use of this form of marking past the withdrawal date. The aircraft had standard 70/71/65 camouflage, with the codes marked in white. The choice of colour for this form of coding appears to have been either black, black thinly outlined in white, or simply white as seen here.

Although of poor quality, this photograph provides a rare view of the aftermath of the March 1938 crisis. The formation of six He 51s, in two groups of three, shows five camouflaged in contemporary fighter colouring of 70/71/65, with the sixth aircraft still in its RLM 63 overall colour. All had post October 1938 proportion Balkenkreuze on the fuselage and Hakenkreuze on the fin. The use of camouflage for this aged fighter biplane indicates its stand-by status during 1938. All have had their wheel spats removed, an addition that served little purpose once the type had left front line service. No markings relating to Fliegerschule identity can be seen, even on the 02-coloured machine.

This Hs 123 A-1 provides an interesting side view for a type that, at least initially, would be deemed by the OKL as obsolescent, only to then be found a continuing role well into the war (mid-1944 for the Hs 123). BQ+NL, W.Nr. 2729, was photographed at an unknown location in the winter of 1939-40, serving at a Stuka Vorschule, polishing up dive bombing and ground-attack skills of pilots before final graduation and posting to operational units. The French campaign was thought to have been the end of the road for the type operationally, yet in June 1941 it would start a new career on the Eastern Front. This example had retained its old pre-1938 61/62/63/65 camouflage to which had been added the 1939-type Stammkennzeichen and revised Balkenkreuz proportions for fuselage and under-wing positions. The type's sesqui-plane configuration required the 'under-wing' Balkenkreuz to be displayed beneath the upper wing, and outboard of the heavy interplane strut assembly, pushing the Balkenkreuz further outboard than laid down in the markings directive of the period. It also meant that Stammkennzeichen did not have each pair split by the Balkenkreuz of the lower surface of the bottom wing – instead the two letters appeared as spaced on the fuselage. It is worth noting that at this time period, aircraft used by the advanced training units did not display any yellow markings to fuselage and wing undersurfaces.

BELOW: Additional markings gradually appeared on aircraft of the advanced training unit, as this Hs 123 A-1 illustrates. Photographed in the summer of 1941 at Stuka Vorschule 1, based at Bad Aibling, W.Nr. 2303 had had a white number added to its Stammkennzeichen identity markings; the white of the '25' presumably signifying its Staffel identity within the school structure. The camouflage was well-worn 70/71/65, patched where the original unit markings had been painted out to apply the Stammkennzeichen letters – all of which indicates that this particular aircraft had been used operationally before being transferred to training duties. Bomb racks beneath the lower wing had also meant that the Stammkennzeichen markings had had to be transferred to the under side of the top wing, inboard of the heavy interplane struts.

TJ+MR, a Ju 88 A of FFS C 14 displays one of the many variations on Stammkennzeichen markings utilised by training schools to identify different formations within the parent unit (see Volume Two, page 394). In this instance the black letters 'TJ' were thinly outlined in yellow with the 'MR' markings in white. If the normal Staffeln/Gruppe system was in use (training units had large holdings of aircraft), it might be interpreted as yellow identifying the 'Gruppe' and white identifying the 'Staffel'. Under high magnification, yellow appears also to have been used around the front, curved edge of the engine cowlings. The unit's Hans Huckebein emblem (a black cartoon-style raven holding a letter C, marked on a yellow disc in the outline of a gimbal mounting) can just be seen below and slightly forward of the cockpit. C-Schulen were used for advanced flying training. In the background is an He 111 with two long, vertical aerials on the fuselage, one just aft of the cockpit, the other just above the Balkenkreuz. Both aircraft have standard 70/71/65 camouflage.

One of the most versatile advanced fighter training aircraft, the Ar 96 B provided the finishing polish to pilots before they moved on to final training with the contemporary single-engine fighter aircraft, some thirteen Jagdschulgeschwadern eventually using the type. W.Nr. 4381 carried the three-number code system used by aircraft engaged in the final stages of fighter pilot training. Insufficient data is available to be specific about the numerical allocation beyond the assumption that the first number identified the 'Staffel' or school and the last two digits probably identified the individual aircraft. Three digit sequences commencing with 2, 4 and 5 have been recorded, and an Ar 96 marked '254' has been identified positively with JFS 2 based at Magdeburg/Zerbst. If correct, the subject aircraft may have been from the same training unit. This system reintroduced a form of distinctive identification markings for advanced flying schools, something lost when Stammkennzeichen were adopted in 1939, in place of the previous school system. The standard colours of white, red and yellow were used to identify 'Staffeln' and Gruppe'-types of divisions with advanced training units. The numerals and spinner on '261' were yellow. Camouflage was 70/71/65, enhanced somewhat by the wet atmosphere, making the overall finish look shiny and darker. (Via Jessen)

The Fw 58 served on all war fronts, usually retaining its standard 70/71/65 camouflage; however, those sent to North Africa usually were camouflaged in some manner to match the changed topography. Initially, some serving in the Mediterranean and North Africa were given Arabesken in 78 colouring over their 70/71 camouflage, but once the desert campaign had begun in earnest, new deliveries of the type were painted in tropical colours of 78 and 79 at point of manufacture, as seen here on KK+UB. This aircraft appears to be in pristine condition, the propeller blades showing no wear at all and the paintwork still had its initial faint glossiness, which soon became a matt surface, indicating that it had seen very few flying hours – probably just the time necessary to fly from its point of manufacture to the Luftwaffe Maintenance Depot where the Arabesken were added over the standard camouflage. That could have been at Erding, on the route south into Italy and on to North Africa (see Volume One, page 88) or any one of a number of bases made available to the Luftwaffe in Italy and Sicily. Enlargement of the original photograph shows no evidence of a splinter pattern beneath the snaking lines, confirming that the upper surface colour was just 79, over which lines of 78 had been applied (see Volume Two, page 346 for an identical scheme on an Fw 189). The Arabesken colour, where it touched, matched the lower surface colour, confirming that the addition had been made using 78-colouring. Note that the spinner also had had lines of 78 added to it – again duplicating the form of scheme seen on the Fw 189 mentioned above. The use of 78 to break up the sand coloured upper surfaces would have been quite effective at the altitudes used by this class of aircraft. The background appears to be typical of North Africa rather than Sicily, so KK+UB had probably just arrived from whichever Luftwaffe Maintenance Depot had undertaken the additional camouflage work.

In September 1945 part of the aircraft dump at Fassberg contained this Ju 87 A, marked SI+AU in black, thinly outlined in white, another example of additions to Stammkennzeichen markings used at some training units. The aircraft is also interesting in that it has retained its pre-war camouflage colours of 61, 62, 63 and 65, in this instance in Scheme A pattern. The Hakenkreuz has been repositioned to the fin, and the Balkenkreuz updated to the October 1939 standard but with the wrong proportions for the wing marking – a common error at that time (see Volume Two, page 436). This was possibly a reserve machine, because regulations applying to refurbishment of camouflage after 1,500 flying hours/two years service in the open would have required some remedial attention – and old stocks of 61, 62 and 63 in the mid-war years were not available. To the right are the remains of an Fw 190 with dark two-tone camouflage along the fuselage spine, probably 81/82. The wings, however, appear to be camouflaged in 76 (it matches the side surface colouring) with 75-coloured segments. This was probably a mismatch of major components at final assembly stage, as wings arriving from sub-manufacturing sources were pre-painted in camouflage. Note that only the Balkenkreuz on both wings (the starboard being more easily seen) had been toned down with an overspray of what appears to be 82. The Balkenkreuz style of the fuselage marking was anomalous, being the old form with black centre and full, black-edged white angles. (D. Vincent)

The last Arado light aircraft to be actively developed for production was the Ar 396. The V1, SO+DA, seen here wore standard 70/71/65 camouflage with the side panel to the engine still in 02. The official camouflage directive was approved for application to the V4 on 5 May 1944, and counter-dated on receipt on 9 May 1944 at the Paris facility of SIPA, where development was conducted on the prototypes. Construction was also initiated at the Letov works in Czechoslovakia, but none were finished in time for Luftwaffe service. Note the 'E-PARIS' and 'V4' stamps. The standard tabular information in the bottom right-hand corner was marked '8-396.000.009', with 'V4' added by hand, and contained the additional handwritten notation "güt auch für A-0" (good also for A-0), showing that the camouflage directive had been also approved for the pre-production series. The pattern was the standard Sichtschutz – Muster A of the type that had superseded the pre-1938 61/62/63/65 form (see Volume One, page 71 – the October 1939 Ju 52 diagram is a very good match). Camouflage colours were identified only by the relevant RLM numbers (70, 71 and 65), with no written colour descriptions added. The note marked beside the colour key was a reminder that the grid squares for each component had to be of equal dimensions. The camouflage pattern had also to be applied in the same manner when marked on the airframe. The date of this drawing is highly significant, within just two months of the 1 July 1944 Sammelmitteilung, notifying the introduction of 81 and 82 for uppersurfaces of such aircraft. This provides clear indication of the swiftness with which the latter colours were introduced after a year's delay, having being promulgated on 21 August 1943 as ready for use (see Volume One, pages 104-105) – an important piece of information, revealing the crisis triggered by the Allied invasion. Retention of 65-colouring did not last long, that colour being separately withdrawn just six weeks later in the cascade of changes to colours which included RLM 83 and the undocumented green-blue colour that appears to have accompanied it.

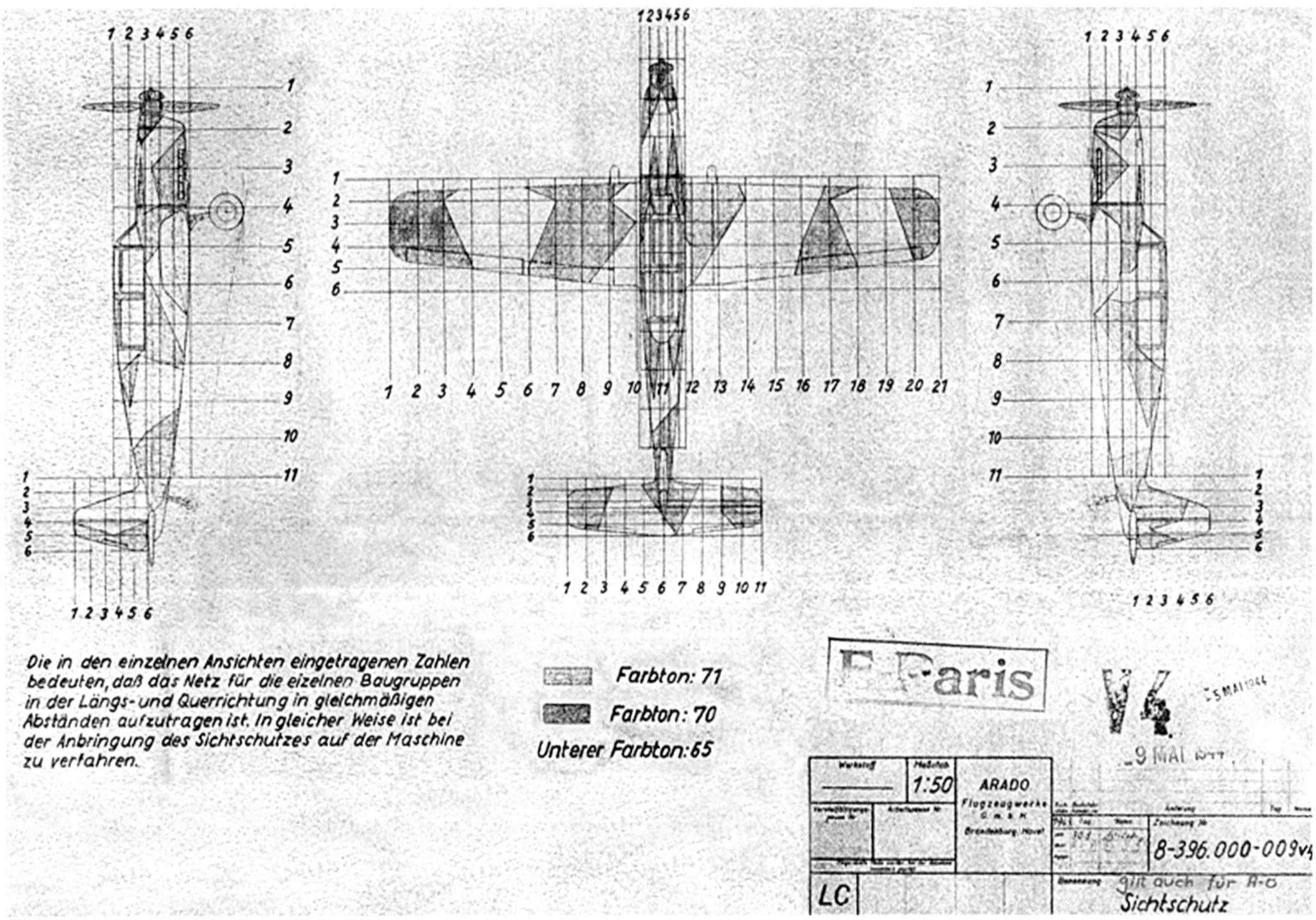

Training aircraft schools often had their own, distinctive unit emblem painted on their aircraft. The choice of subject varied widely from cartoon to coats of arms. Often the coat of arms choice reflected the city/town where the unit had either first formed, or was permanently based. This example is unidentified, but clearly is a coat of arms rather than a made-up emblem. The metal panel on which it had been painted was finished in 70/71/65 indicating that it had been taken from an aircraft equipping one of the advanced flying C-Schullen. Significantly, where some of the camouflage paint had flaked off, the background colour of 02 matches that shown behind the coat of arms, indicating that the aircraft had originally been operated in an 02 finish, then camouflaged at a later date. The coat of arms appears to have been added to the camouflaged finish, as remnants of green paintwork within the marking show. The aircraft type unfortunately has not been positively identified, but this example gives some idea of the colourful nature of the school emblems.

An eclectic collection of training aircraft found by the occupying forces at a large airfield complex, damaged airframes assembled for reconstruction or salvage of parts. They help to illustrate the hugely varied array of markings used by some training schools. The Fw 190s alone show an interesting and confusing array of variation to both their markings and camouflage. On the far right, straddled by the large hoist framework, an Fw 190 marked VO+L3 in black, has late war colouring of 81/82 taken right down the sides of the fuselage. Immediately in front of it is an Fw 190, 'White 6', with extremely pale camouflage colouring to its fuselage, probably 76 overall with just 75 used for the extreme uppersurfaces. The white Balkenkreuz marking was almost lost against the colouring. The wings were replacement items, still in their overall 76-colouring. The discarded original starboard wing shows the 75/76 form of colouring. Of the next two Fw 190s, the one at the back was marked '??+SB' in black, with a white '6' of distinctly different style to the white '6' marked on the previous aircraft; camouflage appears to be 74/75/76. The Fw 190 F in front of it, in the same form of camouflage, has had the last two letters of its black-painted Stammkennzeichen, DO+SO, thinly outlined in yellow. This was possibly W.Nr. ???413 based on the fact that Fw 190 F, 'White 7' (outlined red), DO+SM W.Nr. ???411, was found at Metz-Frescaty airfield, in France. The dark angular patch behind part of the 'SO' marking may have been where a previous marking had been painted out. Next to it is an interesting combination of camouflage. A very thinly proportioned red '15' marking appears to the left of the Balkenkreuz marking, with an enigmatic 'F-8', in black, thinly outlined in yellow, to the right. The fuselage was camouflaged in 81/82 or 81/83, making a distinct contrast with the pale 76-colouring of the lower area of the side surfaces. The wings, in contrast, have had their original pale colouring, (probably 75/76) overpainted with strokes of 82 - possibly a case of replaced wings and an attempt to better harmonise the overall camouflage. (Hidden in the deep shadow, on the extreme left of the photograph, is another Fw 190 on which the same has been done to its previously pale-coloured wings. Next to it is another Fw 190, with wings that had been almost given a monochromatic finish on the outer two-thirds. The Balkenkreuz marking appear to be slightly over-sized.) Of the remaining Fw 190, in the foreground, little is visible other than for its dark late war colouring. It had a Stammkennzeichen marked in black, thinly outlined in white, but it is impossible to determine what it was. The two badly damaged He 111s behind the Fw 190s were in 70/71/65. (S. Coates)

D-1890 was an aircraft with an interesting history. According to the 1930 race listing, only three Messerschmitt M 23b aircraft were entered: 'A1' by Verein f. Luftfahrt of Gladbach; 'B3' by the Deutsche Verkehrsfliegerschule G.m.b.H at Braunschweig; and 'D7' by the Leichtflugzeugklub München e. V. However, 'F1' (D-1890), fitted with a Siemens Halske Sh 13a radial engine, clearly also was a Messerschmitt M 23b, one of the eight machines entered by the parent company. The anomaly attracted comment by the foreign press. When first registered in July 1930, it was recorded (incorrectly) as an M 23c and the race organisers probably used the civil register to construct their race programme. In March 1931 Rudolf Hess bought the aircraft. It then passed to Akademische Fliegergruppe TH at Hannover in April 1932, and finally to the DVL e.V in Berlin in March 1934. It was removed from the civil register in September 1936. Overall colour was cream with black civil registration and the race number sited beneath the cockpit and the usual weights and loading table marked just forward of the empennage. The Stanavo Oil Company emblem – looking so much like a stylised letter 'F', and easy to confuse with the Fieseler Company logo which sloped in the other direction – can be seen on the engine cowling, while the BFW Company logo of a stylised eagle was marked on the rudder. Not usually noticed was that the company logo also included the letters BFW in block lettering marked vertically between 'wing tip and tail feather'. The colour possibly was red.

Exhibiting distinctly more colour was the Arado Ar 11a, D-1875 'D1', one of three entered by the parent company Arado Handelsgesellschaft m.b.H of Berlin, the other two carrying the race numbers 'C8' and 'C9'. Colour was cream with black trim applied with some style to fuselage and wings and tail surfaces, the latter two having stylish scalloped edging to the colour division. The Arado Company emblem was marked at the tip of the rudder. Wheel hubs were also in cream. The lower section of the engine cowling had been removed for work on the engine at the time the photograph was taken. First registered in July 1930, it was sold after the racing season to the RDL Erprobungs-Stelle at Travemünde, surviving until June 1937.

Produced in relatively small numbers, the elegant Ar 79 served in communications duties in Luftwaffe service. Pre-war, the type was produced in an overall cream colour with black civil registrations, as seen here on D-ECGQ and D-EADC. The cream colouring was probably matched to one of the RAL-system colours such as RAL 1014 Elfenbein or RAL 9001 cremeweiß, but no doubt having an Arado company designation. The RAL colour standard was used by all, but written colour designations were very much the prerogative of individual companies, making identification impossible without access to original company painting schedules and their link to the German colours standards of the RAL system.

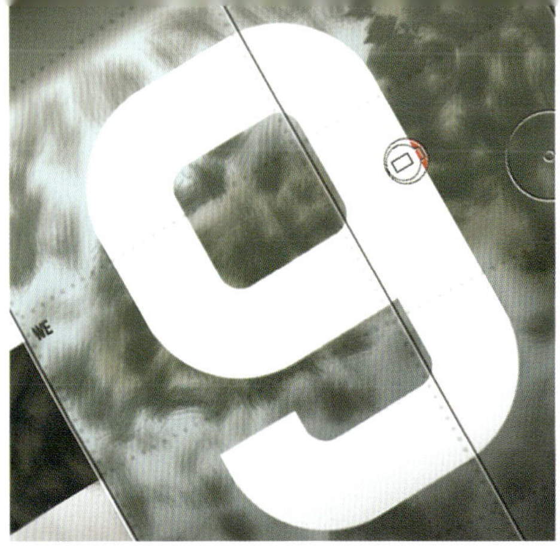

Instruments and Cockpit Interiors

The Ju 52 was one of a number of civil aircraft which served with Luft Hansa in the 1930s. Instrumentation remained the same when it entered military service. Apart from the radio equipment, which was upgraded to military standard, very little else was changed in the cockpit layout, other than the internal colouring. Flight instruments were situated in front of each pilot's position, with engine instrumentation and engine controls placed centrally. (Lufthansa)

The cockpit layout in the civil He 111 C was, again, almost identical to that of the military version which would soon appear. The second pilot doubled as the flight engineer/radio operator and had a plug-in control column provided for emergency flying. Engine instrumentation was clustered in front of his position, only the first pilot having a direct view of the flying instrumentation. Instrument panel colouring at this time was Wermalin-Spritzlack grau matt (a sprayed matt pale grey finish). (Lufthansa)

The military version of the He 111 exhibits relatively little change to the cockpit layout, the glazed nose section, with its forward facing machine gun and bomb aiming position, introduced the bulkhead door on the right with its removable post hole-style circular hatch. The engine instruments on the right-hand side remained as before, minus the two engine temperature gauges. The centre console above the boost levers and throttles had the temperature gauge and chronometer moved to make way for two boost gauges. On the flying instruments panel in front of the pilot, the fine and course altimeter had been moved to the left-hand side, replacing the compass selector, and substituted with a combined course indicator/turn and bank/heading repeater unit.

By the time the He 111 P appeared, the nose section had undergone significant redesign, but the instrument types remained the same. The repositioned instrument panel still held the same selection of flying instruments for the pilot with engine instrumentation to the right. Note that by this time, the military practice of marking each instrument with a letter/number combination (to identify, for maintenance purposes, its connection via the electrical system) had become standard (see Volume One, page 207). (H. Mankau)

LEFT: The major change to the military version of the Ju 86 was relocation of the equipment on the right to provide access to the revised and glazed nose section. The engine controls – throttles/fuel levers/radiator flap controls – attached to the base of the instrument panel had their housing shape changed to blend into the revised access passage. The instrument panel itself had been replaced with a flat version, reduced in overall height for better forward vision, cropped on the starboard side to bring all the instrumentation into a better position for the single pilot operation; but for the most part instruments simply had been rearranged. The same dual engine instrumentation, e.g. tachometers, fuel flow, fuel pressure, oil pressure, temperature, etc. used the same instruments as in the civil version. The pilot's flying instrumentation – turn and bank, variometer, altimeter, artificial horizon, course indicator, air speed indicator – had been shifted around somewhat. Some instruments were updated versions, but overall they remained the same. (L. Tabori)

The D 18, which joined Luft Hansa in 1936, provided generous space for its crew positions, even in its civil configuration. The radio operator/navigator's position also was well equipped. In the left-hand view the radio equipment was attached to the bulkhead. Just above the edge of the table was the radio compass equipment (marked PIELEN in white) with the switches for the compass and radio selection and the large dial which allowed radio bearings to be located. The box and single switch below the porthole was the Lorenz beam approach selector. The right-hand view gives more detail of the space and equipment layout. The trailing aerial control with its crank handle was mounted above the Morse key with an altimeter, voltmeter, amp meter and a selector on the sloping panel facing it. On the end bulkhead was more Lorenz equipment. For its time this was a very sophisticated layout. The interior of the aircraft itself was painted with silver, a good anti-corrosive measure used on flying boats of the period. (Lufthansa)

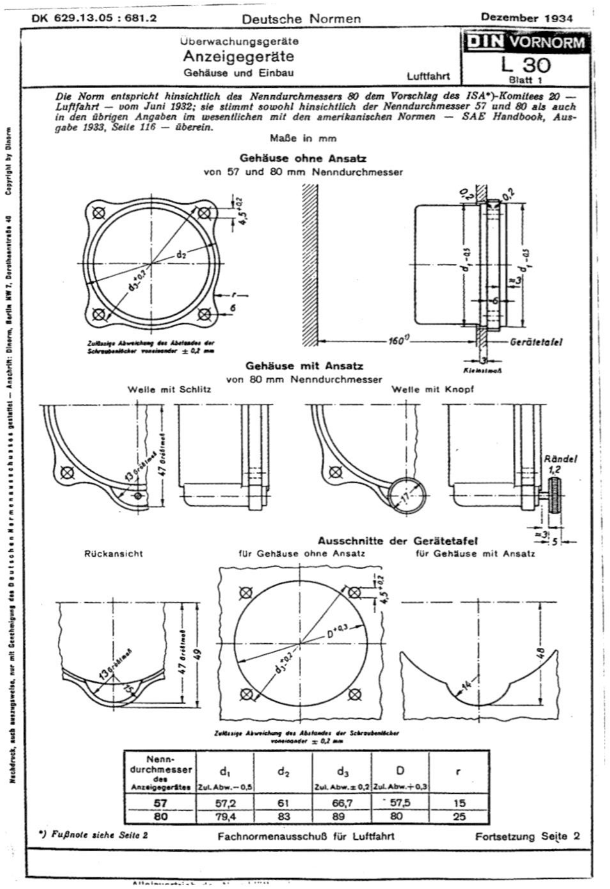

DK 629.13.05 : 681.2 Deutsche Normen Dezember 1934

Überwachungsgeräte
Anzeigegeräte
Gehäuse und Einbau — Luftfahrt

DIN VORNORM L 30 Blatt 1

Die Norm entspricht hinsichtlich des Nenndurchmessers 80 dem Vorschlag des ISA)-Komitees 20 — Luftfahrt — vom Juni 1932; sie stimmt sowohl hinsichtlich der Nenndurchmesser 57 und 80 als auch in den übrigen Angaben im wesentlichen mit den amerikanischen Normen — SAE Handbook, Ausgabe 1933, Seite 116 — überein.*

Maße in mm

Gehäuse ohne Ansatz
von 57 und 80 mm Nenndurchmesser

Zulässige Abweichung des Abstandes der Schraubenlöcher voneinander ± 0,2 mm

Gerätetafel

Kleinstmaß

Gehäuse mit Ansatz
von 80 mm Nenndurchmesser

Welle mit Schlitz — Welle mit Knopf

Rändel 1,2

Rückansicht

Ausschnitte der Gerätetafel
für Gehäuse ohne Ansatz — für Gehäuse mit Ansatz

Zulässige Abweichung des Abstandes der Schraubenlöcher voneinander ± 0,2 mm

Nenndurchmesser des Anzeigegerätes	d_1 Zul. Abw. − 0,5	d_2	d_3 Zul. Abw. ± 0,2	D Zul. Abw. + 0,3	r
57	57,2	61	66,7	57,5	15
80	79,4	83	89	80	25

*) Fußnote siehe Seite 2 — Fachnormenausschuß für Luftfahrt — Fortsetzung Seite 2

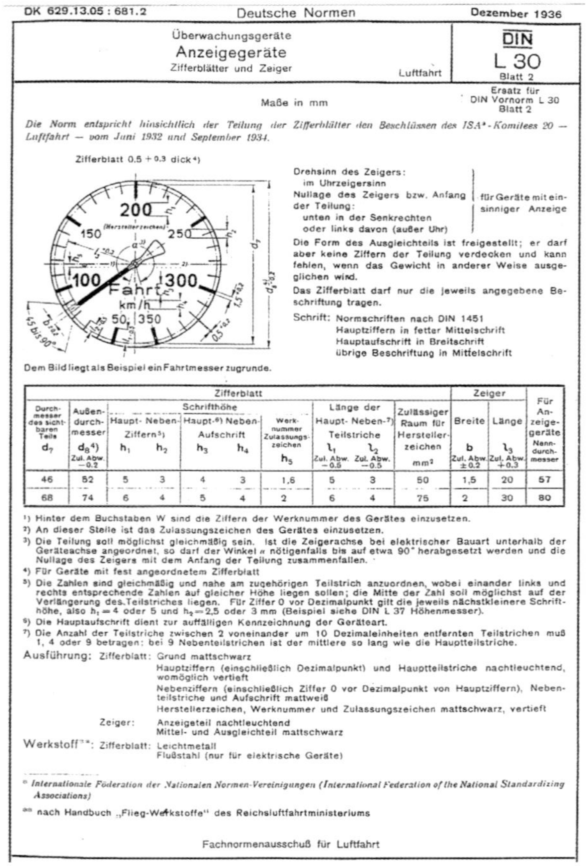

DK 629.13.05 : 681.2 Deutsche Normen Dezember 1936

Überwachungsgeräte
Anzeigegeräte
Zifferblätter und Zeiger — Luftfahrt

DIN L 30 Blatt 2

Maße in mm — Ersatz für DIN Vornorm L 30 Blatt 2

Die Norm entspricht hinsichtlich der Teilung der Zifferblätter den Beschlüssen des ISA-Komitees 20 — Luftfahrt — vom Juni 1932 und September 1934.*

Zifferblatt 0.5 + 0.3 dick[4])

Drehsinn des Zeigers:
im Uhrzeigersinn
Nullage des Zeigers bzw. Anfang der Teilung:
unten in der Senkrechten
oder links davon (außer Uhr)
— für Geräte mit einsinniger Anzeige

Die Form des Ausgleichteils ist freigestellt; er darf aber keine Ziffern der Teilung verdecken und kann fehlen, wenn das Gewicht in anderer Weise ausgeglichen wird.

Das Zifferblatt darf nur die jeweils angegebene Beschriftung tragen.

Schrift: Normschriften nach DIN 1451
Hauptziffern in fetter Mittelschrift
Hauptaufschrift in Breitschrift
übrige Beschriftung in Mittelschrift

Dem Bild liegt als Beispiel ein Fahrtmesser zugrunde.

Zifferblatt: Durchmesser des sichtbaren Teils d_7	Außendurchmesser d_8[4]) Zul. Abw. −0.2	Schrifthöhe Haupt-Ziffern[5]) h_1	Neben-Ziffern h_2	Haupt-[6]) Aufschrift h_3	Neben-Aufschrift h_4	Werknummer Zulassungszeichen h_5	Länge der Haupt-Teilstriche l_1 Zul. Abw. −0.5	Neben-[7]) Teilstriche l_2 Zul. Abw. −0.5	Zulässiger Raum für Herstellerzeichen mm²	Zeiger: Breite b Zul. Abw. ±0.2	Länge l_3 Zul. Abw. +0.3	Für Anzeigegeräte Nenndurchmesser
46	52	5	3	4	3	1,6	5	3	50	1,5	20	57
68	74	6	4	5	4	2	6	4	75	2	30	80

1) Hinter dem Buchstaben W sind die Ziffern der Werknummer des Gerätes einzusetzen.
2) An dieser Stelle ist das Zulassungszeichen des Gerätes einzusetzen.
3) Die Teilung soll möglichst gleichmäßig sein. Ist die Zeigerachse bei elektrischer Bauart unterhalb der Geräteachse angeordnet, so darf der Winkel α nötigenfalls bis auf etwa 90° herabgesetzt werden und die Nullage des Zeigers mit dem Anfang der Teilung zusammenfallen.
4) Für Geräte mit fest angeordnetem Zifferblatt
5) Die Zahlen sind gleichmäßig und nahe am zugehörigen Teilstrich anzuordnen, wobei einander links und rechts entsprechende Zahlen auf gleicher Höhe liegen sollen; die Mitte der Zahl soll möglichst auf der Verlängerung des Teilstriches liegen. Für Ziffer 0 vor Dezimalpunkt gilt die jeweils nächstkleinere Schrifthöhe, also $h_1 = 4$ oder 5 und $h_2 = 2{,}5$ oder 3 mm (Beispiel siehe DIN L 37 Höhenmesser).
6) Die Hauptaufschrift dient zur auffälligen Kennzeichnung der Geräteart.
7) Die Anzahl der Teilstriche zwischen 2 voneinander um 10 Dezimaleinheiten entfernten Teilstrichen muß 1, 4 oder 9 betragen; bei 9 Nebenteilstrichen ist der mittlere so lang wie die Hauptteilstriche.

Ausführung: Zifferblatt: Grund mattschwarz
Hauptziffern (einschließlich Dezimalpunkt) und Hauptteilstriche nachtleuchtend, womöglich vertieft
Nebenziffern (einschließlich Ziffer 0 vor Dezimalpunkt von Hauptziffern), Nebenteilstriche und Aufschrift mattweiß
Herstellerzeichen, Werknummer und Zulassungszeichen mattschwarz, vertieft
Zeiger: Anzeigeteil nachtleuchtend
Mittel- und Ausgleichteil mattschwarz

Werkstoff**: Zifferblatt: Leichtmetall
Flußstahl (nur für elektrische Geräte)

* *Internationale Föderation der Nationalen Normen-Vereinigungen (International Federation of the National Standardizing Associations)*

** nach Handbuch „Flieg-Werkstoffe" des Reichsluftfahrtministeriums

Fachnormenausschuß für Luftfahrt

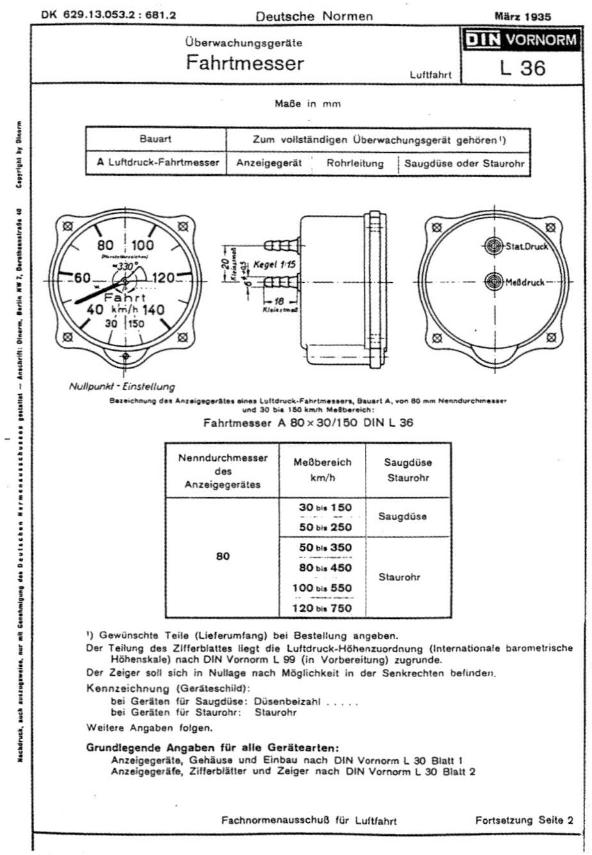

DK 629.13.053.2 : 681.2 Deutsche Normen März 1935

Überwachungsgeräte
Fahrtmesser — Luftfahrt

DIN VORNORM L 36

Maße in mm

Bauart	Zum vollständigen Überwachungsgerät gehören[1])		
A Luftdruck-Fahrtmesser	Anzeigegerät	Rohrleitung	Saugdüse oder Staurohr

Kegel 1:15 — Kleinstmaß — Stat. Druck — Meßdruck

Nullpunkt-Einstellung

Bezeichnung des Anzeigegerätes eines Luftdruck-Fahrtmessers, Bauart A, von 80 mm Nenndurchmesser und 30 bis 150 km/h Meßbereich:

Fahrtmesser A 80 × 30/150 DIN L 36

Nenndurchmesser des Anzeigegerätes	Meßbereich km/h	Saugdüse Staurohr
80	30 bis 150	Saugdüse
	50 bis 250	
	50 bis 350	Staurohr
	80 bis 450	
	100 bis 550	
	120 bis 750	

1) Gewünschte Teile (Lieferumfang) bei Bestellung angeben.

Der Teilung des Zifferblattes liegt die Luftdruck-Höhenzuordnung (Internationale barometrische Höhenskale) nach DIN Vornorm L 99 (in Vorbereitung) zugrunde.

Der Zeiger soll sich in Nullage nach Möglichkeit in der Senkrechten befinden.

Kennzeichnung (Geräteschild):
bei Geräten für Saugdüse: Düsenbeizahl .....
bei Geräten für Staurohr: Staurohr

Weitere Angaben folgen.

Grundlegende Angaben für alle Gerätearten:
Anzeigegeräte, Gehäuse und Einbau nach DIN Vornorm L 30 Blatt 1
Anzeigegeräte, Zifferblätter und Zeiger nach DIN Vornorm L 30 Blatt 2

Fachnormenausschuß für Luftfahrt — Fortsetzung Seite 2

Instrument design was strictly regulated. The DIN (Deutsch Normen) standards authority, which regulated almost everything in the industrial world of Germany, issued the relevant documentation. The December 1936 issue of DIN L 30 incorporated and upgraded the regulatory requirements set by the ISA (International Federation of National Standards Associations) Komitees 20 - Luftfahrt of June 1932. This ran to three pages, but the first two will provide an understanding of how the regulatory process was set out.

The second page provided a specific example, in fact an upgrading, of the September 1934 DIN design specification of the standard air speed indicator. This incorporated all previous requirements relating to dimensions, materials, markings, etc, but now specified the DIN 1451 standard for the various markings, as well as the colour of the instrument face (mattschwarz), with mattweiß for the markings, and the engraving of the manufacturer's name, instrument type, etc. The third page gave data standards for the glass face, edge bevels, etc.

Some instruments were designed for multiple use, incorporating a number of interchangeable dials, with different markings for differing ranges of performance, e.g. air speed indicator, tachometer, variometer, etc. DIN L 36 issued in March 1935 contained two pages, the first setting out the standards required for the instrument.

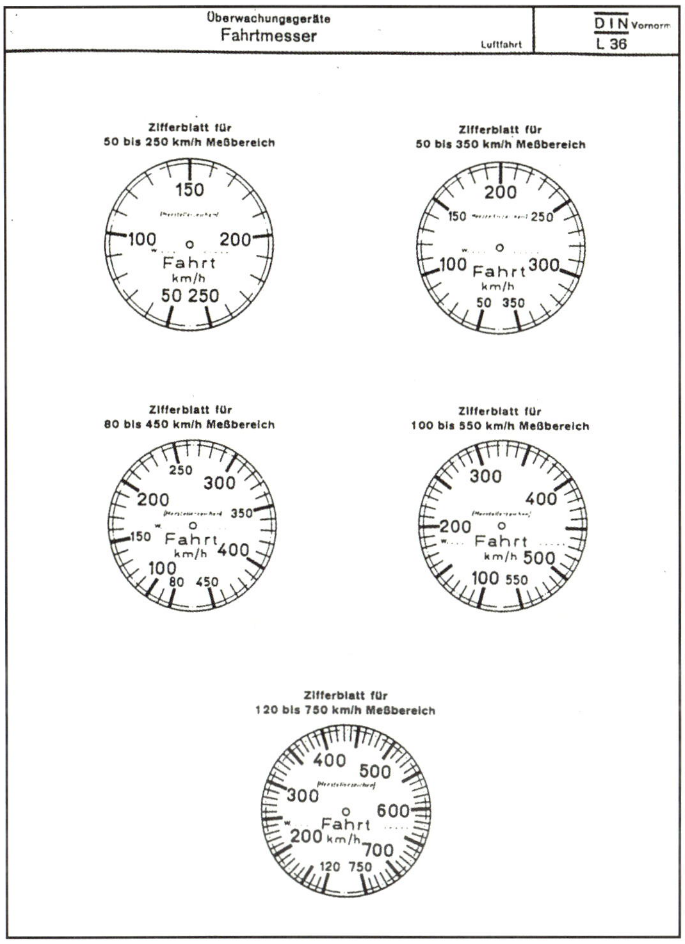

The second page contained five alternative dials for the instrument, graduated from 50 to 250 km/h, 50 to 350, 80 to 450, 100 to 550 and 120 to 750. The other main instrument that required a similarly wide range of dial choices was the fine and course altimeter, five sets of two dials (course and fine graduations) being set down in the March 1935 DIN L 37 document.

An Askania-built example of an Air Speed Indicator calibrated 0-450 km/h from a Patin autopilot system. This was used on some transport aircraft. (P. Cohausz)

Another example, this time from Bruhn – calibrated 0-550 km/h, it was standard fitting for a range of types, e.g., He 111, Hs 126, Hs 129, Ar 76. This specific instrument had a red on white painted marking at the 250 km/h mark, indicating a critical speed limitation for whichever type it was fitted in. (P. Cohausz)

Among the other pre-war aircraft that would serve with the Luftwaffe was the Fw 200. The instrumentation shown here would remain almost unaltered when the type took on its war role. The very long range of this aircraft required two pilots, with duplication of the flying instruments in front of each pilot. (Lufthansa)

At the outbreak of war the Blohm und Voss Ha 139 filled a gap, albeit briefly due to lack of spares, in the Luftwaffe's immediate need for long-range aircraft. Again, little would have changed with its instrumentation. Most instrument types seen here can be found in the earlier photographs.

This view from the flight engineer's station of a Blohm und Voss Ha 139, immediately aft of the cockpit, gives some indication of the sheer interior size of these machines. Transfer of the engine instrumentation to this location reduced the amount of instruments which, for twin-engine aircraft, were required on the main panel in the cockpit. (Lufthansa)

Immediately aft of the pilot in the Ju 87 B, was the radio equipment, circuit breaker panel, oxygen bottle for the rear gunner and ancillary items. Note the bare metal for the floor section and wall – counter to the internal painting standard of the day. (R. Lutz)

The Ju 87 B entered service in 1938, but again its instrumentation consisted mainly of the usual range of well-designed instruments already in use for a number of years in the civil aviation field (see Volume One, page 208). Interestingly, none of the identifying colours had been added to the rim of any instrument, nor had the instrument maintenance system of letter/number symbols been added to the instrument panel, indicating that this was probably one of the early B model machines. By 1938, internal colouring for military aircraft was RLM 02, although the type, but not colour, of lacquer was changed that same year (see Volume One, page 207). Instrument panels, however, were finished in 7107.66. (R. Lutz)

The hand-painted white lettering on each ammunition drum of the Ju 87 B can more easily be seen here, with 'No 1' canister at top left, 'No 2' below it, 'No 5' at the start of the top rack, 'No 11' starting the bottom row, and 'No 7' in place on the machine gun. Interestingly the painter had used 'No', with a white dash below the '0', rather than the more usual German marking style of 'Nr.' Note the manufacturer's data plate rivetted to the cockpit sill, to the left of the machine gun. The optical distortion through the curvature of the side panels is interesting and would not have helped in sighting aircraft in the distance in that region. (R. Lutz)

ABOVE: While the Ju 87 B machine gun retention locks and the cross frame below the gun were painted in 02, the rest of the cockpit was bare metal other than for the gunner's seat and the inside of the canopy framework. The doppeltrommel (double drum) ammunition boxes for the rear gun were black, and each one here has been neatly marked in white with an individual number. The harness for the gunner was natural canvas colour with an 'R' marked on the right-hand lap strap. (R. Lutz)

RIGHT: Interesting variations could occur to standard instrumentation positioning, for example in this Bf 109 test aircraft, for a specific type, when airframes were used for a variety of test purposes. Apart from the large test instrument in the upper right side of the instrument panel, the layout was a mixture of sub-models. The main panel was almost the same as that of a C or D model; however, the fine and course altimeter had been moved up tight against the air speed indicator and the prop pitch indicator had been moved to a position immediately above it (replacing the variometer), freeing-up the upper right quadrant of the instrument panel. The sub-panel contained the type of dimmer switch fitted to the E-1 and E-3 variants, but also present was a ZSK bomb selector panel of the type fitted to the E-4 B and E-7 variants. The 04 yellow painted fuel flow pipe, with its clear glass section, was another feature relevant to late model Es. Note also the two pipes that parallel it and pass through a hole in the instrument panel.

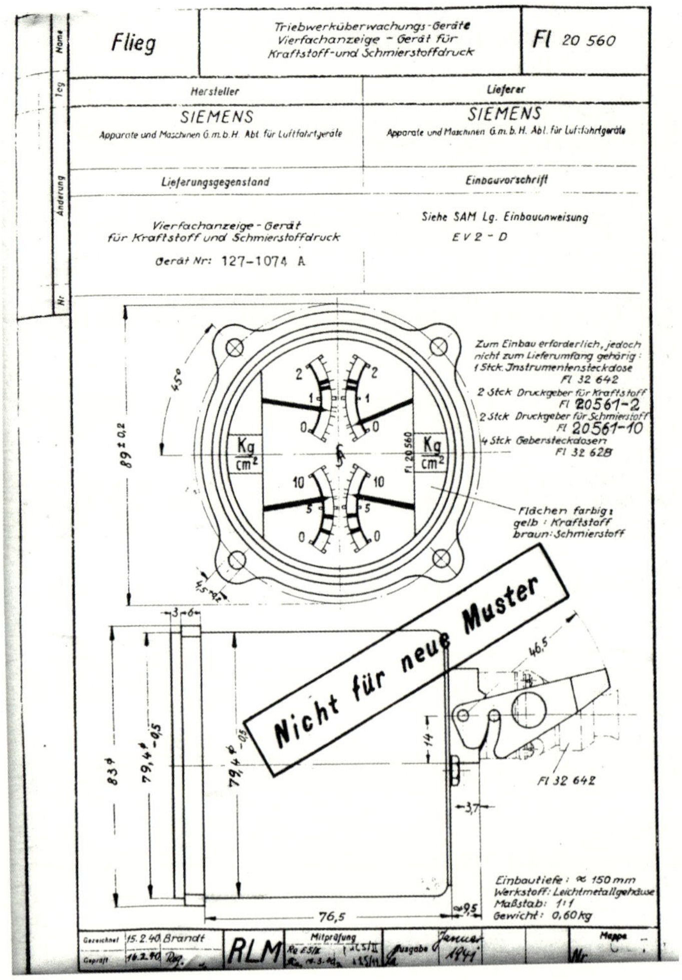

While the civil DIN standards still applied to instrument design, the RLM issued its own specifications to industry when it wanted revision to existing instruments or new types of instruments. This sample was issued to Siemens with the specifics set down. This double instrument (fuel and oil pressure) was for twin-engine aircraft (hence the four sets of readings) and had been first prepared in February 1940 with final approval granted in January 1941. It laid down all dimensional aspects and instrument face design, and included the Luftwaffe stores identification number (Geräte nr: 127-1074 A) and its 'FL' (Fliegluft) number, (FL 20 560). The latter was further expanded by notations shown to the right of the main drawing to cover variations to the instrument's use as a pressure reading instrument. Note also the inclusion of colour coding instructions for the instrument face sections: gelb; kraftstoff (yellow for fuel); brauns chmierstoff (brown for oil). The over-stamped marking Nicht für neue Muster (Not for new type) was added at a later stage, and related to requirements of new aircraft types, probably with a different pressure range. (P. Cohausz)

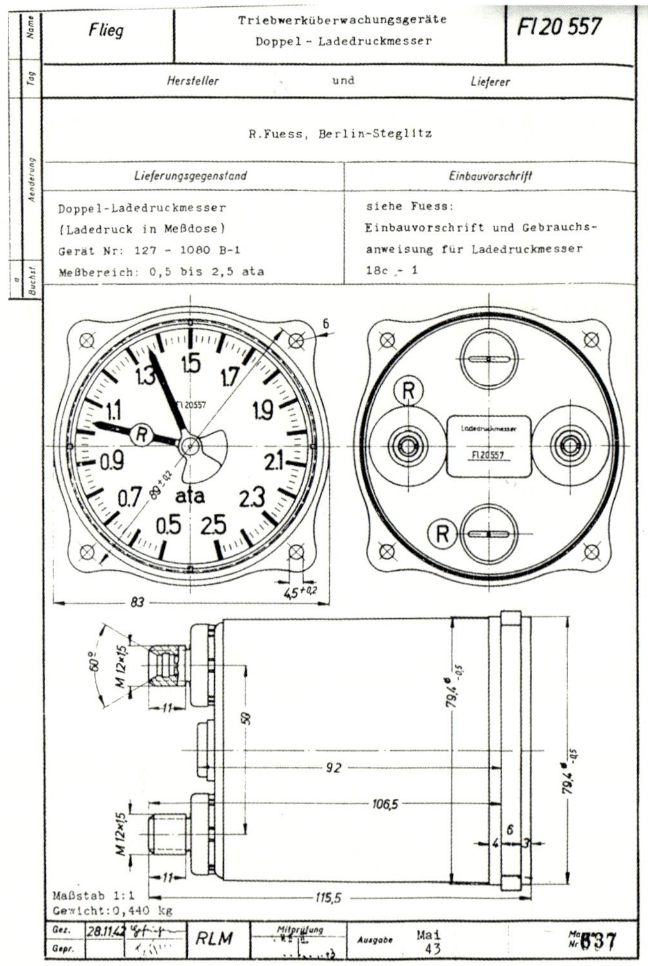

This example of a design specification issued to R. Fuess, Berlin-Steglitz for manufacture, was for a double reading manifold pressure gauge. First drawn up in November 1942, FL20 557 was approved in May 1943. This was an updated version of FL20 550, which had a lower pressure range. (P. Cohausz)

This is the earlier form of the same instrument, Lpa-3A, produced by the Askania company and used pre-war on such types as the Ju 87 A and B. It continued in use until around 1939/1940. The instrument face is engraved with the identification number FL 20 550 on the right, its construction number on the left and the Askania emblem in the upper half. The revised version seen in the illustration above, had the FL number placed centrally above the pivot point for the needles. Note the significant reading points, marked in red on white background strips, such additions being made once the instrument was fitted to a specific aircraft type. (P. Cohausz)

The Junkers W 33 was obsolete by the mid-1930s but it continued in a training capacity into the war years. Given its age, the instrumentation seen here illustrates that little changed in terms of the types of instruments other than updating in some instances for higher ratings. Compare the instruments shown here with those for the W 34 hi, shown in the following diagram below. (J. Vasco)

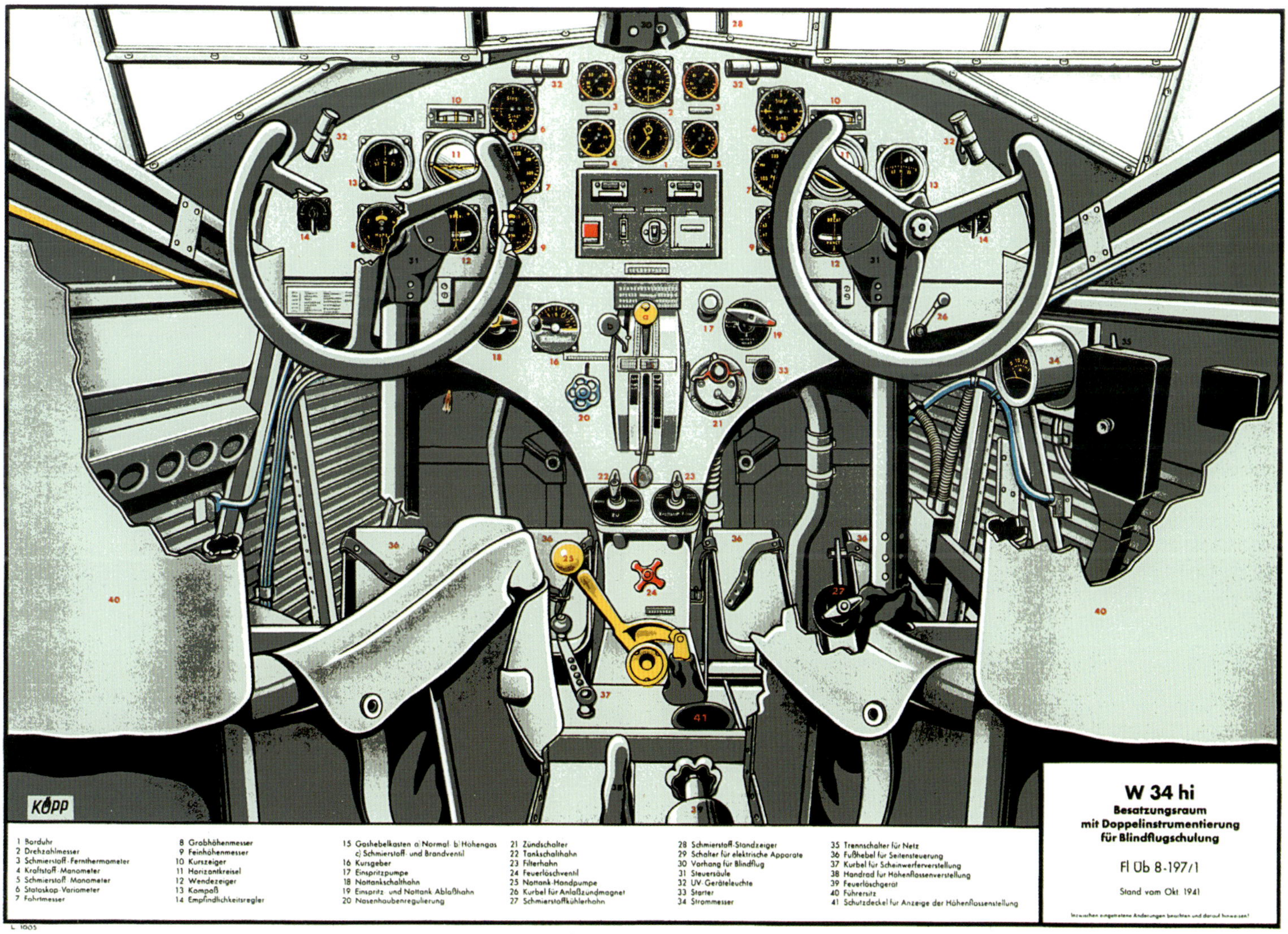

The main difference is to the central instrumentation arrangement, produced by the change from an in-line engine to a radial engine. (K. Kössler)

In the early stages of the night bombing raids against England, the Luftwaffe employed the 'Knickebein' system of blind bombing for its target marking force. Using 30 mega-cycle Lorenz beam approach equipment, the aircraft travelled down the main beam in the course of which it would cross three intersecting beams, the first 30 km from target. At the second intersection, 15 km from the target, the navigator pressed the start button on a large clock-like device fitted to the starboard side of the He 111. As the aircraft intersected the third cross-beam, 5 km from the target, the navigator pressed a button to stop the first hand of the clock, leaving the second hand to continue its sweep. As the two hands of the bombing clock coincided, the bombs were automatically released. This rare shot of the navigator (Erich Sommer, later to become a celebrated pilot and Ar 234 commander in Italy) in one of KG 100's He 111s shows part of the bombing clock device. (E. Sommer)

Ju 52s were fitted out for a range of training purposes – navigation, radio, etc. This example had a bench fitted in front of the instrumentation on the starboard side. The spartan interior shows the 02-painted metal skinning, seat and table frames. Only areas visible through glazing were required to have a finish of 66. (H. Mankau)

The Siebel Fh 104 was an elegant five-seat light transport aircraft that gained a good reputation in pre-war competitions. The cockpit design for the single pilot was exceptionally clean and efficient, as can be seen here. Instrumentation was composed of standard types. Cockpit colouring appears to be a light grey.

LEFT AND BOTTOM LEFT: Production of the Fh 104 A ceased in May 1942, only 46 machines having been constructed, most of them used by the Luftwaffe for communications and liaison duties. The company had already commenced design of an enlarged development of the type, designated Si 204, which first flew in 1941. This diagram of its instrument layout gives a good idea of the instruments used. (K. Kössler)

ABOVE: The large Patin KT-p2 direction finding repeater compass, set in front of the second pilot. (P. Cohausz)

1 Notwendezeiger
2 Nahkompaß
3 Gerätebrett
4 Blendschirm mit Plexibeleuchtung
5 Geräteleuchte
*6 Kurszeiger
7 Funknavigationsanzeige für Blindlandung
8 Funknavigationsanzeige
9 Fahrtmesser
10 Wendehorizont
11 Variometer
12 Verdunkler
13 Feinhöhenmesser
*14 Kurskreisel (be Nichteinbau der Kurssteuerung Kurbel zur Basisverstelung von Führertochterkompaß)
15 Führertochterkompaß
16 Kippwechselschalter für Scheinwerfer
17 Kippschalter für UV-Leuchte
18 Entlüftungshahn für stat. Druck
19 Selbstschalter mit Fernauslösung für Bordnetz
20 Grobhöhenmesser
*21 Hauptschalter für Kurssteuerung
22 Netzausschalter
23 Dreifachtrimmanzeige
24 Ablenkungstafel für Nahkompaß
25 Ablenkungstafel für Peiltochterkompaß
26 Elt. Ferndrehzahlanzeige links
27 Elt. Ferndrehzahlanzeige rechts
28 Ladedruckanzeige ilnks
29 Ladedruckanzeige rechts
30 Vierfachanzeigegerät links
31 Vierfachanzeigegerät rechts
32 Elt. Kraftstoffvorratsanzeige links
33 Elt. Kraftstoffvorratsanzeige rechts
34 Kippschalter für Kraftstoff-Außenbehälterpumpe links
35 Kippschalter für Kraftstoff-Außenbehälterpumpe rechts
36 Merkleuchte für Kraftstoff-Reststandswarnung links
37 Merkleuchte für Kraftstoff-Reststandswarnung rechts
38 2-Meßstellenumschalter für Kraftstoff-Vorratsmessung links
39 2-Meßstellenumschalter für Kraftstoff-Vorratsmessung rechts
40 Deviationstafelschieber mit Ablenkungstafel für Führertochterkompaß
41 Stromspannungsmesser
42 Außenlufttemperaturanzeige
43 Temperaturanzeige für Flügelnasenheizung
44 2-Meßstellenumschalter für Flügelnasen-Temperaturanzeige
45 Peiltochterkompaß
46 Peilumschalter
47 Rahmendrehschalter
48 Höhenmesser
49 Fahrmesser
50 Zielflugpeilempfänger
51 Senderschaltkasten
52 Schwingungsanzeige
53 Kippausschalter für Anodenumformer
54 12-Lampengerät für Landeklappen und Fahrwerk
55 Flachzündschalter
56 Kommandoschalter für Verstell luftschraube
57 Gashebel
58 Schnellstophebel
59 Feststellvorrichtung für Gas
60 Brandhahnhebel
61 Dreifachdruckknopfschalter für Landeklappenverstellung
62 Dreiachsentrimmschalter
63 Kippwechselschalter für Fahrwerk
64 Anlaßschalter links
65 Anlaßschalter rechts
66 Druckmesser für Feuerlöscher
67 Hahnbatterie für Feuerlöscher (Motor links)
68 Hahnbatterie für Feuerlöscher (Motor rechts)
*69 Notauslösung für Kurssteuerung
70 Notauslösung für Fahrwerk
71 Gerätebock
72 BB.-Fußhebelpaar für Seitensteuerung
73 BB.-Steuersäule für Höhensteuerung
74 BB.-Handgriffe für Quersteuerung
*75 Richtungsgeber
76 Borduhr
77 Borduhrleuchte
78 StB.-Fußhebelpaar für Seitensteuerung
79 StB.-Steuersäule für Höhensteuerung
80 StB.-Handgriffe für Quersteuerung
81 Borduhr
82 Borduhrleuchte
83 Spant 4
84 Betriebsdatentafel
85 Spant 5
86 Telefon-Anschlußdose ADb 11
87 Kurbel für Schiebefenster-Betätigung
88 Schiebefenster
89 Bediengriffe für Klappfenster
90 Klappfenster
91 UV-Leuchte
92 Blendschutzvorhang
93 Notaussteigklappe
94 Taste
95 Verkleidungskanal für elt. Leitungen
96 Verteiler für Funkanlage
97 Leuchtpistole
98 Ausschußöffnung für Leuchtpistole
99 Bediengriff für Steuerentkupplung
100 Handpumpe für Landeklappen-Notbetätigung
101 Flugzeugführersitz
102 Funkersitz

Die mit * versehenen Geräte sind nur bei Einbau der Kurssteuerung vorhanden

Si 204 A-0 und A-1
Besatzungsraum
Blick nach vorn
Fl Üb 8-207 6
Stand vom März 1943
Inzwischen eingetretene Änderungen beachten und darauf hinweisen!

The interior of the Si 204 could be reconfigured for a number of roles. This one was used for navigation training and had normal seats fitted at the front of the cabin, with a table and second seat for two pupils at a time at the rear of the cabin. The bottles below the table, marked with two white and one blue band, were oxygen bottles. Interior colouring was 02. (H. Mankau)

This original blind flying panel taken from an Si 204 D shows the small variation of this later model, the large combined course indicator and master compass repeater having been removed and the Patin repeater compass moved into the central position. Colour was RLM 66. (P. Cohausz)

A rare example of a blind flying instrument sub-panel from the very late war period. Constructed from wood (in places the red wood primer coat can be seen where the RLM 66 finish has worn away), use of non-strategic materials and the high calibration of the ASI, clearly indicate that it came from an aircraft which was not piston-engine powered. In size and shape it is similar to blind flying sub-panels installed in the Me 163 B, Do 335 A-0, Me 262 A, and BV 155 B-1. The Do 335 and BV 155 panels, like this one, had had the 'oxygen on'- disc indicator (below the ASI) blanked off with a strip of thin metal, rounded on the left-hand end. However, the arrangement and selection of instrumentation differed from all those types. Starting top left, this panel has an Air Speed Indicator (calibrated to 1,000 km/h), a combined Artificial Horizon/Turn and Bank indicator, and a Fine and Course altimeter. Below, from the left, are a second Fine and Course altimeter, a Variometer, and an AFN-2 Indicator (radio navigation). Inclusion of two identical types of altimeter creates a mystery. Also the lack of a compass on the panel is intriguing, possibly indicating a short range, the AFN-2 indicator providing necessary navigational information. (An oil-filled magnetic stand-by compass was usually provided in all aircraft types, for safety, but mounted separately in the cockpit.) It is possible that the panel may be from one of the few Me 263 airframes that were built and flown, several examples of which survived to the end of the war. (B. Sartwell)

Errata and Addenda to **Luftwaffe Camouflage and Markings 1933-1945** Volumes One and Two by K.A. Merrick.

I am most grateful to a number of people who have either contacted me direct or posted comments on the 12 O'Clock High site, identifying errors in captions to both Volume One and Two, listed below, as well as providing additional information. These include further corrections to Volume One, received after Volume Two had gone to press.

Ken Merrick
May 2006.

ERRATA – Volume One

Chapter 5
Page 86 - Lower caption. The Ju 88 identified is not the one recovered from the sea (Ju 88 A-5 W.Nr. 0886146) but instead Ju 88 C-2 W.Nr. 0881033 recovered from a Norwegian mountain site, where it had remained exposed to extreme weather conditions for around 50 years. The reference to the effectiveness of the anodised internal finish remains just as pertinent.

Chapter 7
Page 159 - Bottom photograph caption. Type identification should read 'Bf 109 G-6/R6', not 'G-2/R6'. The Werknummer stated is also incorrect (see Addendum).

ADDENDA – Volume One

Chapter 7
Page 156 - Bottom photograph. Caption should include the aircraft's Werknummer of 10639. This is the aircraft now displayed at the RAF Museum, Hendon.

Page 177 - Lower photograph caption. This is not a Bf 109 G-6 as stated, but a Bf 109 G-14, W.Nr. 413601, 'Black 7', the first of this sub-type to fall into Allied hands.

Appendix A
Page 200 - Paragraph two; line 8. After "....63 and 65 lower surfaces." add, "(In all, Sweden eventually would receive 40 Ju 86s from Germany)."

Page 200 - Paragraph two, line 12. After "... the Swedish order." add, "The Hungarian order was eventually increased to 66, 63 of which are known to have been delivered as bombers, with the remaining three possibly supplied in transport configuration. Austria received three Ju 86s, but all were returned to Junkers' Dessau plant before the Anchluß."

Page 202 - last paragraph. "In 1942, Germany also honoured an export agreement with Turkey for the supply of 75 Fw 190 A-3 aircraft, which were delivered in standard Luftwaffe camouflage of the period."

ERRATA – Volume Two

Introduction.
Page 230 - Paragraph Two - From line 7, commencing "Included with this volume are..." text is a repeat of the opening of following paragraph.

Chapter 8
Page 233 - Photograph caption, delete "was" after "....03+F(or E)20, ..."

Chapter 9
Page 237 - Top caption, "-" missing after word "staff".

Chapter 10
Page 251 - Top caption, should read in part, "...'Yellow 4' marking identifying it as 9. Staffel, III. Gruppe...". Further, reference to the aircraft in the foreground as "..the Gruppen Adjutant's aircraft..." is ambiguous. It was not intended to imply that it was the Gruppen Adjutant's aircraft of the III. Gruppe mentioned above, but rather 'a Gruppen Adjutant's aircraft', as indicated by the lack of 'sine wave' form of III. Gruppe marking. (See Addenda below for further details). The Werknummer quoted for the latter aircraft (10681) is incorrect, the most probable one being 10501.

Page 252 - Top right caption. The aircraft is not from an 11. Staffel as stated. The unit has been identified as I./Erg.KG(J) based at Pilsen and Budweis.

Chapter 11
Page 262 - Lower right-hand caption - delete last sentence "The purpose of the white edged circle is unknown". (See caption to photograph, Chapter 10, page 252 upper left.)

Page 296 - Bottom caption. This is a "Fw 1900 F-8, W.Nr. 582304, 'Yellow 13', (not 'Yellow 11' as stated) photographed at Neubiberg.

Page 301 - Bottom caption, "letters" should read "markings".

Chapter 12
Page 305 - Lower photograph caption should read "the black overall finish applied to the sides and lower surfaces (uppersurfaces retained their original two-tone camouflage) with the..."

Page 310 - Upper caption - this should read "Ju 88 G-6", not "Ju 88 G-7a".

Page 310 - Left-hand caption, centre - this should read "Ju 88 G-6", not "Ju 88 G-7a". Line 2, delete comma after "was" in "...again was, 76..."

Page 311 - Bottom caption, line 1, insert "of NJG 1" after "..." this Ju 88 C-6" (not "Ju 88 G")

Chapter 13
Page 319 - Bottom caption, as printed, applied to two photographs of this aircraft, the photograph inadvertently deleted from the book showing the code "6X" marked in white outline form as noted in the existing caption. The "X" can just be discerned in the existing photograph, in the shadow of a crew member standing on the ground. See this volume for second photograph.

Chapter 17
Page 361 - Captions to diagrams at bottom of page are reversed.

Chapter 17
Page 370 - Last paragraph. Reference the comment, quote, *"Photographs show the same four-letter code, TS+MA, applied to several aircraft, Bf 109s and Fw 190s..."* is incorrect. The TS+MA codes appeared only on two Fw 190s, not on any Bf 109s.
Page 364 - Lower caption should read " Bf 109 G-4 trop aircraft..." not "G-6 trop aircraft..."

Appendix F

Page 402 - Bottom left caption, the location is Milovice, not Prague-Letnany.

Page 408 - source credit to top photograph missing, "…..via F. Grabowski."

Appendix G

Page 427 - Bottom caption, line 2, "coded" should read "code"

Appendix H

Page 433 - top right photograph, aircraft type misidentified as "Ju 88". Delete existing caption and replace with the following;
"Me 410, Werknummer 130378, demonstrates the very low position used on this type for marking of the Werknummer, the location being common to most Messerschmitt fighter aircraft. However, the size of the numerals as shown here, was larger than those normally seen on the type. The relatively low positioning of the Hakenkreuz was another characteristic of the Me 410 tail markings."

Appendix I

Page 443 - Lower middle caption, should read "….Bf 109 E-7/b trop of 7./ZG 1….." not "Bf 109 F trop".

Page 446 - This is not an Fw 190 G-8; correct identity is Fw 190 F-8, W.Nr. 584592, 'Black 14', photographed at Neubiberg.

ADDENDA – Volume Two

Chapter 10

Page 249 - Top caption. This is an Fw 190 F-9.

Page 251 - Top caption. Additional information provided has identified the location as possibly Bir-el-Abd, in which case the photograph was taken after the Luftwaffe had abandoned the site in November 1942. The Bf 109 G-2 <3 in the foreground belonged to Stab./JG 77. While the quality of the original photograph does not allow absolute certainty, the Werknummer most likely is 10501, not 10681 as cited. The Bf 109 F-4 trop, 'Yellow 4', belonged to 9./JG 27. It has been pointed out that III./JG 27 was the only Bf 109-equipped unit in North Africa to use this form of III. Gruppe identification, thus confirming the unit identity.

Chapter 11

Page 296 - Top caption. Bf 109 G-14, possibly W.Nr. 464456, of II./JG 52 photographed at Neubiberg.

Page 296 - Middle caption. Bf 109 G-14s of II./JG 52 photographed at Neubiberg.

Chapter 13

Page 318 - Top caption. This is possibly Ju 87 G-2, W.Nr. 494200 from the base at Pilsen.

Chapter 14

Page 330 - Bottom caption. The location is Hradec Králové (Königgratz)

Chapter 19

Page 388 - Bottom caption. The aircraft probably belonged to FFS C16.
Page 396 - The caption refers to the use of the *"…white on black 'F' emblem…"*. This has now been identified as the STANAVO company emblem (see below), the well known US aviation fuel company, and comprised a stylised emblem of a plan view of an aircraft, set at an angle, with the company name STANAVO cutting the uppermost portion of the wings (but almost impossible to detect in any of the photographs). See caption on page 130 of this book, top photograph.